THEORIES OF IMAGE FORMATION

THEORIES OF
IMAGE
FORMATION

Edited by **DAVID F. MARKS**

University of Otago
Department of Psychology
Dunedin, New Zealand

BRANDON HOUSE   **NEW YORK**

PUBLISHED BY
BRANDON HOUSE, INC.
555 RIVERDALE STATION
NEW YORK, NEW YORK 10471

FOR ORDERS, WRITE TO: P.O. BOX 240
BRONX, NEW YORK 10471

LIBRARY OF CONGRESS CATALOG CARD NUMBER: 84-072473
INTERNATIONAL STANDARD BOOK NUMBER: 0-913412-18-X
MANUFACTURED IN THE UNITED STATES OF AMERICA

CONTENTS

CONTRIBUTORS

AKHTER AHSEN, Editor, *Journal of Mental Imagery,* P.O. Box 240, Bronx, New York

B. R. BUGELSKI, Department of Psychology, State University of New York at Buffalo, 4230 Ridge Lea Road, Buffalo, New York

MICHAEL C. CORBALLIS, Department of Psychology, University of Auckland, Private Bag, Auckland, New Zealand

GEIR KAUFMANN, Department of Cognitive Psychology, University of Bergen, Bergen, Norway

STEPHEN M. KOSSLYN, Department of Psychology and Social Relations, Harvard University, William James Hall, 33 Kirkland Street, Cambridge, Massachusetts

DAVID F. MARKS, Department of Psychology, University of Otago, P.O. Box 56, Dunedin, New Zealand

ELIOTT MORDKOWITZ, Department of Psychology and Social Relations, Harvard University, William James Hall, 33 Kirkland Street, Cambridge, Massachusetts

PETER MCKELLAR, Department of Psychology, University of Otago, P.O. Box 56, Dunedin, New Zealand

GOSAKU NARUSE, Dean of the Faculty of Education, Kyushu University, Hakozaki, Higashi-Ku, Fukuoka, Japan

JAMES D. ROTH, Department of Psychology and Social Relations, Harvard University, William James Hall, 33 Kirkland Street, Cambridge, Massachusetts

JOHN C. YUILLE, Department of Psychology, University of British Columbia, Vancouver, British Columbia, Canada

ACKNOWLEDGMENTS

The following authors and publishers are acknowledged for granting permission to reproduce figures:

Chapter 6, Figure 3: L.A. Cooper and R.N. Shepard. Chronometric studies of the rotation of mental images. In W.G. Chase (Ed.) *Visual information processing*. New York: Academic Press, 1973.

Chapter 8, Figures 9-11: A. Paivio. Images, propositions, and knowledge. In J.M. Nicholas (Ed.) *Images, perception, and knowledge*. Dordrecht: D. Reidel, 1976.

Chapter 8, Figures 12-13: A. Paivio. Comparison of mental clocks. *Journal of Experimental Psychology*: *Human Perception & Performance, 4,* 61-71. Copyright 1978 by the American Psychological Association.

Chapter 8, Figures 14-15: S.M. Kosslyn. Scanning visual images: Some structural implications. *Perception & Psychophysics, 14,* 90-94. Copyright 1973 by the Psychonomic Society.

Chapter 8, Figures 16-17: S.M. Kosslyn, T.M. Ball, and B.J. Reiser. Visual images preserve metric spatial information: Evidence from studies of image scanning. *Journal of Experimental Psychology*: *Human Perception & Performance, 4,* 47-60. Copyright 1978 by the American Psychological Association.

FOREWORD

The return of an interest in imagery by psychologists reflects a decline in the restraints of behaviorism against the phenomena of consciousness. At the same time it would be wrong to call this simply a return to an introspective psychology of consciousness. Behaviorism enriched scientific psychology by extending its boundaries in many directions at the same time that it restricted it in other directions. The new cognitive revolution that accelerated in the 1950's and 1960's did indeed bring back attention to neglected areas but it remained largely in the hands of those who had not lost their respect for behaviorism's quantitative precision and its use of instruments of measurement. The tachistoscope and the computer may have largely displaced the maze, classical and operant conditioning apparatus, and memory drums, but the logic of mainstream psychology did not change greatly.

The revival of imagery came in the early 1970's after the new cognitive psychology was well along, with the controversies over imagery sparked by Pylyshyn's review in 1973. His rejection of the image as representational of, or analogous to perceptual experience (for example, "pictures in the mind") in favor of an abstract linguistic or propositional interpretation sparked a controversy that has persisted ever since between what Dennett called the *iconophiles* (who liked the representational position) and the *iconophobes* (who were against it, in favor of something else).

Many significant contributions have been made by those using the methods evolved in the behavioristic era and in the early years of cognitive science, even though heated disputes remained among the authors accepting this general approach. The approach is well represented in this book that derives from the First International Imagery Conference held in New Zealand in 1983, edited by David Marks, well qualified by his own empirical work.

There was another camp represented, however, not following a single line, but in general critical of the more conventional tie to experimental and cognitive psychology. Akhter Ahsen, in the first chapter, supplies a very scholarly historical background, and is one of those skeptical about the positivistic approaches that neglect the richness of imagery. His own approach is by way of a Triple Code Model of imagery and imagination. The three components may appear in any order, but none can be neglected: (I) imagery of imagination; (S) somatic response, including feeling and emotion; and (M) meaning, including cognition or semantic relevance. His half-dozen books on eidetic psychotherapy and related imagery studies began to appear in 1965, before the current interest in imagery had accelerated, so that his impatience with some of the restrictiveness in the new developments is understandable. He fears, for example, that emphasis on experimental design may end "with no homage paid to the real stalk or content of the experience itself." An extreme is John Yuille's expression of his anti-experimental orientation: "The practice of experimental psychology with the control which the methodology provides the experimenter, means that any reasonable hypothesis can be confirmed in the laboratory." (To me this lies outside my experience, but perhaps he would classify the hypotheses I have had to reject as unreasonable ones.) There are many critics of the positivistic approach in psychology and although this statement is an extreme there is something to be said.

The varieties of viewpoints expressed, and the evidence of excitement and strong commitments are representative of much of psychology today. In all areas psychology is in flux, much self-searching is going on, and the right to question established beliefs and practices is widely acclaimed.

The processes of assimilation and accommodation described by James Mark Baldwin and later by Jean Piaget well characterize imagery theories of the varied kinds described in this useful collection of papers. All agree that imagery must be assimilated within contemporary psychology. There is uncertainty where the accommodation shall lie. Shall those interested in imagery accommodate their theories and practices to the prevailing cognitive psychology, or does psychology have to accommodate by becoming an entirely different science to be at home with the metaphorical and mythological aspects that lie at the heart of experiences of imagery and imagination?

The search for answers will haunt the careful reader of this book, who, at another level, will have gained much knowledge relevant to the answers.

ERNEST R. HILGARD
Stanford University

INTRODUCTION

The occasion for this volume was set by the First International Imagery Conference held in Queenstown, New Zealand in December, 1983. Participants from sixteen countries came to this meeting held beside beautiful Lake Wakatipu and the majestic Remarkables mountain range. This was certainly no ordinary meeting. The professional backgrounds of the participants were diverse: art, drama, literature, music, neuropsychology, nursing, physical education, psychology, psychotherapy, religious studies, and social work. Topics on the program spanned an equally broad range. Amidst all the diversity of special interests we shared a single common concern: the image, its theory, method, and application.

This volume publishes six Queenstown keynote addresses, in original or revised form, presented by Akhter Ahsen (U.S.A.), Michael Corballis (New Zealand), Geir Kaufmann (Norway), Gosaku Naruse (Japan), John Yuille (Canada) and David Marks (New Zealand). Another keynote speaker, Peter McKellar (New Zealand), prepared a chapter especially for this volume, and Richard Bugelski (U.S.A.) and Stephen Kosslyn (U.S.A.) who were unable to attend the Queenstown Conference have contributed chapters representing their seminal work on the theory of image formation. The final chapter was prepared after the conference and represents my own analysis of how some of the unresolved problems concerning the nature and function of image formation might eventually be solved.

Many different issues and themes emerge in this volume. None of the contributors displays a frozen, static representation of past thinking but, instead, we are treated to a new generation of ideas. Many new possibilities and vistas are opened, as though a new vision of psychology is being shaped like a veritable Phoenix.

The psychology of the late twentieth and early twenty-first centuries will hopefully be a radically different one. The existing schools have had little to say about the nature and function of *image experience*, in fact, about consciousness in general. The currently accepted paradigms and methods of investigation plainly are inappropriate to the task. The neo-behavioristic cognitive psychology shares the anti-introspectionist stance of its behaviorist predecessors. Methods which may be natural and effective for laboratory research with rats and pigeons are clearly not appropriate for research with humans. The best efforts of cognitive researchers to study mental mechanisms have used verbal learning or response time measures. The former confounds imagery with language effects and the latter remain ambiguous in the absence of some converging operations which of necessity in imagery research must be introspective.

Akhter Ahsen provides an illuminating analysis of the history of the empirical method. Positivism, operationalism and other forms of excessive empiricism which exclude experience from the science of psychology become themselves a kind of metaphysics. We are given a systematic analysis of reflection which goes beyond the somewhat narrow interpretation allowed by that great British empiricist, John Locke. Fiction, metaphor and myth are shown to provide a link between literature and scientific discovery through the generation of new ideas and images. Dr. Ahsen explains his tripartite or Triple Code Model of image formation, and provides new insights into the role of vividness, unvividness, and dissociation in the empirical investigation of individual image sequences. Particularly fascinating is the light he throws on the vividness paradox.

Akhter Ahsen's analysis mirrors several issues raised elsewhere in this volume. For example, John Yuille's chapter shows how the psychologist's laboratory experiment has been turned into a tool, not of discovery, but of confirmation. The investigator's ideas on how imagery processes might operate have been quite literally imposed, albeit unwittingly, upon the subject's overt performance via the contrived experimental context. This is an ironic but nevertheless fatal turn of events for committed operationalists but it is inevitable if one believes that observations can never be the product of the observer's theoretical assumptions.

Peter McKellar outlines the philosophical, historical and literary sources of thinking about the unconscious, and shows how many imagistic and dissociational phenomena require an unconscious associative mechanism which might only become conscious through various kinds of therapeutic or transformational maneuvers. The role of analogy and metaphor in scientific theory has a parallel in the insightful awareness of associative links and relations which may occur in imagery therapy. Just as Akhter Ahsen has shown us how myth, metaphor and fact are inseparably linked in the magical thinking of the human psyche, Peter McKellar illustrates how theories of personality are themselves constituted in metaphorical and analogical terms. Only by actively attempting to translate these from one system to another, in the manner of a good therapist, can illumination replace fixation.

The role of imagery in learning (and vice versa) are considered in some detail by Gosaku Naruse and Richard Bugelski. Considering that the former is a Japanese hypnosis researcher in the Buddhist tradition and the latter is a North American verbal learning researcher in the behaviorist tradition, the meeting of minds is extraordinarily close, especially when it comes down to the theoretical nitty-gritty. Over three decades, Naruse's subtle and careful researches on the mental conditions for imaginal conditioning have opened doors of discovery which should never again be closed. The hypnotic state is unquestionably suited to the empirical exploration of imagery. With some notable exceptions, Western psychologists have generally disparaged the hypnotic state as a research tool, on much the same grounds that imagery experience has itself been tabooed. This gap in our knowledge does not appear in Eastern psychology, where the self-induced hypnotic state is more widely acknowledged as a useful skill.

Unlike the Western emphasis on the specialness of such a state, Naruse believes that all can become proficient hypnotic subjects following appropriate instruction. Yet we find the same potential in our own concepts of what hypnosis might be like. Consider, for example, the idea that being hypnotized is similar to our absorption in an enjoyable or exciting play, film, or story. Who does not have the capacity to be so absorbed? The laboratory context must again be examined and perhaps will be found wanting in its ability to inhibit or facilitate hypnotic experience depending upon social and personality factors interacting within a given individual. But this says nothing about the general capacity for experiencing an altered state of consciousness such as hypnosis which must indeed be universal.

Scales such as the *Stanford Hypnotic Susceptibility Scale*, Wilson and Barber's *Creative Imagination Scale*, and my own *Vividness of Visual Imagery Questionnaire* all give a wide range of scores and reliably so if the conditions of testing are not varied. But if these conditions are changed, and different contexts and instructions are provided, low scorers can become high scorers and/or vice versa. Like all measurement, the obtained scores depend upon the context. This is as true for psychology as it is for physics or chemistry. Dick Bugelski argues for a single Pavlovian conditioning model of learning. Images are given the central linking role between S and S, or S and R, and the neural processes responsible for learning are held to be the same as the ones underlying the images. Bugelski explains "instrumental conditioning" as a special kind of Pavlovian conditioning in which neural, imagistic processes correlated with the proprioceptive stimuli accompanying reinforcement, antedate the operant response. In other words, the animal presses the lever because it *images* the reinforcement which will follow. Reinforcement has thus been returned to its common-sense role as a *motivator* which is what most people really thought it was all along. What a pity a Dick Bugelski wasn't in business when Watson took psychology into the "rat phase" of the "Dark Ages" and again when Skinner launched the second "pigeon phase."

Stephen Kosslyn, James Roth and Eliott Mordkowitz outline their computational theory of image generation. This kind of theory is based not upon the computer as metaphor but on the computer as model. As Kosslyn and his coauthors freely admit, mimicking the mind on a computer is a tricky business, and there are numerous alternative architectures for any particular mental task. A major challenge for this vigorous new approach is how to decide which solution is the most "correct," or if indeed there is a way of doing so.

Michael Corballis describes some elegant experiments he and his colleagues have conducted on the discrimination of letters from their mirror-images when presented in differing orientations. A process of "mental rotation" had previously been attributed to the fact that the subjects' response times are linear with the angle of the test stimulus (Cooper & Shepard, 1973). Mike Corballis mentions that the subjects themselves often have difficulty reporting on the characteristics of the presumed rotation and suggests that the impression of rotation may be a rationalization rather than "a true experience." There is no doubt that introspective awareness of images of rotating objects can occur but this may not always be so under experimental conditions. Is it helpful to include both cases under the same rubric?

Geir Kaufmann articulates his doubts concerning the conceptual difficulties of cognitive imagery models which he divides into three classes: analog (Kosslyn, Paivio, and Shepard), propositional (Anderson and Bower, Chase and Clark, and Pylyshyn), and anticipational (Neisser). None of these emerges unscathed by serious conceptual problems. Kaufmann presents his own theory of image formation which sees imagery as an "ancillary representational system operating mainly within and under conceptual control of the linguistic symbolic system." However,

Kaufmann does allow in his scheme thinking without language, in the form of "action programs tied to and activated in more specific action contexts" and also "some rudimentary and primitive pre-linguistic imagery." Kaufmann's willingness to entertain images of varying genesis and function is a welcomed step, although pejorative labeling creates conceptual problems of another kind (see Akhter Ahsen's chapter, Section II-1).

In Chapter 9, I outline some recent ideas and data on the neuropsychological basis of imagery. This evidence does not appear to be in tune with the relegation of imagery to ancillary roles. On the contrary, there are strong grounds for thinking that while imagery without language may be possible, language without imagery may not be. Recent EEG experiments suggest that introspective awareness of mental activity requires cortical activation at the tertiary temporo-parieto-occipital junction in the left hemisphere. This region is activated in vivid and non-vivid imagers during calculation (of which both groups of subjects are introspectively aware) but in vivid imagers only during visualization (Marks, 1985). These results confirm the doubts raised in Chapter 9 that imagery is a purely right hemispheric function and suggest that regions in both hemispheres are involved in different components of the image-generation system.

The final chapter begins with my personal outline of systems within psychology which have attempted either to account for or ignore imagery. These systems include what I have pessimistically called psychology's "Dark Ages." The darkness descended, I believe, when consciousness, all its contents, and the associated introspective methodology were ignobly ejected. A new era of awareness of the potential of rigorous research on human consciousness seems to be dawning. Imagery, and the processes associated with its formation, understanding, change, and application, provide a natural starting point for such investigation. It is to be hoped that this volume provides some inspiration to others who feel moved toward the empirical investigation of imagery as an exemplary conscious activity.

This project would not have been possible without the help of many people. In Queenstown itself, John Baird, Oliver Davidson, Marcel Farry and Michael Whitbrock provided much needed assistance. Myrna Wooding, Margaret Gilkison, and Helen Mathieson assisted in the preparation of the edited manuscript. Research funding from the University Grants Committee, the University of Otago Research Committee, and the Cancer Research Society is gratefully acknowledged.

My intellectual debts are fairly diverse and I have been helped by many who cannot all be named. Undoubtedly the most important influences I am aware of were my parents, Mary and Victor Marks, my brother Jonathan, and the following teachers, colleagues and friends: Akhter Ahsen, Jack Clarkson, Peter McKellar, M.D. Vernon and George Wallace.

References
Cooper, L.A., & Shepard, R.N. (1973). Chronometric studies of the rotation of mental images. In W.G. Chase (Ed.), *Visual information processing*. New York: Academic Press.
Marks, D.F. (1985). Imagery, consciousness, and the brain. In R.G. Russell, D.F. Marks, & J.T.E. Richardson (Eds.), *Imagery 2*. Dunedin: Human Performance Associates.

DAVID F. MARKS
Dunedin, New Zealand

CHAPTER **1**

Image Psychology and the Empirical Method

Akhter Ahsen

The return of imagery in psychology was ironically predictable in the very failure of behaviorism to account for the origin of its data in experience. Behaviorism had designed for itself a self-imposed limitation in the study of phenomena, especially when we find in the history of thought a consistent tendency to return to the notion of *experience* in one form or another. This notion exists in all philosophers, empiricists, strict positivists and even dyed-in-the-wool behaviorists, although they may not admit it openly. However, the notion of experience which appears so

This chapter is reprinted from the *Journal of Mental Imagery* (1985), a special theme issue, "Imagery and Scientific Methodology."

1

basic and fundamental is not as simple to put into practice. There have been all kinds of views concerning the concept of experience in the history of philosophical and psychological thought which makes the notion most baffling. Luckily, however, complexity of an idea is its very history, which also feeds its growth. The science of psychology comprises the history of notions concerning the terms which have evolved in it, in which sense the notion of experience and its close association with imagery has been the very essence of scientific psychology. Imagery happens to be the most basic in psychology and is the hidden ghost in most of its changing paradigms.

However, the concept of imagery in general psychology is a concept with special problems of its own. To be scientific one must start on an empirical basis concerning the image by provisionally rejecting what Rene' Descartes called the "natural attitude." The statement that our commonsense information about images is generally not reliable is thus the first postulate in imagery theory. But, luckily or unluckily, the image is a paradox, since it is a something, yet it is not a thing in the usual sense of the word because it exists only in the head. No one can, following Johnson's famous argument, kick the image like one would kick a stone and prove that it does exist. But look closely; even Johnson did not go outdoors to kick a stone; he only informed about it. At the empirical level, the stone, the toe, the kick, and the pain, along with the proof, are all mental images. If the science of psychology has to deal with a real subject matter such as imagery, it has to treat the mental image as real in the same way as all things are real at the mental level, and then accept or reject, in the next step, the commonsense view of it. Thus, one starts with the postulate of a "real thing" called "image" at the introspective plane about which the "natural attitude" may hold an erroneous view which needs to be examined and modified through scientific observation. Somehow, this step was lost in the stampede as empiricism developed its historic perspectives on the image in the psychological science. As a result, the introspective field was not treated as real.

I. HISTORY OF EMPIRICISM

1. Locke's Empiricism Revisited

The empirical philosophy which developed in England in the seventeenth century and is at the source of psychology's scientific character was started by Thomas Hobbes (1588-1679) but its truly successful founder was John Locke (1632-1704). Locke laid the foundation of empiricism by asserting that experience is and should be the only means of attaining knowledge, that all the materials of reason and knowledge come from experience, and that in the sphere of experience alone all our knowledge is founded and from that it ultimately derives itself.

To underline the ultimate importance of experience, Locke rejected Descartes' earlier notion of "innate ideas," replacing it with the venerable old scholastic *tabula rasa* concept expressed in Thomas Aquinas' famous axiom: There cannot be anything in the intellect which was not previously in some form in the senses. Locke stated that the human mind is like a "white sheet," that all the ideas of the human mind are acquired by the *senses* or through *reflection*, which was defined as a "perception of the operations of our own mind." He defined mind as playing the passive role of a receptacle, place or compartment in which impressions were stored. Thus, Locke rejected Descartes' notion of "innate ideas" and replaced it with a more thoroughgoing empiricism in which knowledge evolved not from metaphysical principles but through psychological analysis of experience. From this hard-core empiricism evolved experimental psychology which earned for Locke the deserved title of "founder of experimental psychology." Experiment in a way is rooted in the demonstration of experience. The word "empiricism" or "empiric" is from the Greek word *empeiria*, meaning "trial, attempt, experiment." Unfortunately, the current *empeirikos* or "experimental" in modern experimental psychology carries less of Locke's original empiricism or even the engaging flavor of the Greek root from which it is derived. This loss can be directly attributed to a certain trace of ambiguity in Lockean thought itself.

Locke had made his departure from Descartes' notion of innate ideas not because that notion was not true (later day science has, in fact, implicitly rejected Locke's *tabula rasa* in the discovery of DNA), but rather because its metaphysical background stood in the way of building knowledge out of pure experience itself. Thus, Locke's notion of the *tabula rasa* was a strategic notion very much in the tradition of Descartes' own dualism rather than a metaphysical statement about existence. In fact, empiricism, whatever its implied strategy, cannot, by the rules, make an ultimate statement on the nature of existence itself, only on how the description of a fact present in the context of this existence must be approached: experientially. Thus, to draw any pure metaphysics out of empiricism itself is to defeat the very aim of empiricism, which is to serve as a tool for the growth of science. Many later forms of empiricism, such as behaviorism and some of the experimental psychology of today, have been guilty of turning this excellent tool into a kind of pseudo-metaphysics.

Deficiencies in modern experimental psychology can be directly traced to this unfortunate twisting of goals of empiricism and its historical intentions. This twist created pseudo-metaphysical anti-experiential trends in current experimental methodologies, changing them at times into form without content. Clearly the care and the search for content progressively diminishes as the experimental form increases its hold. This inverse correspondence between method and truth leads to a victory of the form. The final outcome is emphasis on the elegance of the experimental design with no homage paid to the real stalk or content of the experience itself. In the end the experimental method fawns its own pretentions in the shape of fake content, as an artefact, as pointed out by Yuille (Chapter 8). Surely, as Berkowitz and Donnerstein (1982) point out, most human activity is context dependent, field based and, by nature, active. Yuille (1985) in the light of this reached a conclusion, stating: "The practice of experimental psychology, coupled with the control which the method ology provides the experimenter, means that any reasonable hypothesis about human cognition can be confirmed in the laboratory." With a sympathetic tone, and without accusing his opponents of deliberate fraud, Yuille assumes a very strong posture toward researchers in modern psychology especially Paivio, Shepard, Finke, Pinker, and Kosslyn, etc., making a plea for a new empirical departure in psychology toward paradigms which do not exhibit the effects of task demands, but involve context-dependent and field-based research.   He points out that "This thesis is not anti-empirical, only anti-experimental." To Yuille, excessive control and precision paradoxically lead to loss of meaning and value in research paradigms. The history of negative control on data or experience is long and goes all the way back to Plato.

2. Control of Sensation: Defamation or Grace?

Plato has rightly been accused for being the first critic of sensation, but he has also been shown to be truly cognizant of its existence. To him, sensation was a befouler of pure knowledge, since true knowledge could only be gathered from pure intellect. Plato recognized that sensation was material and he considered that to be its main blemish. He also considered sensation to have a supreme *power* in this world because it made the pure intellect suffer by imposing a *delusionary* state on it. For Plato, sensation was an empirical fact, but not a philosophically desirable goal.

For the scholastic philosopher Thomas Aquinas, who proposed *tabula rasa*, everything in the intellect came previously in some form from the senses, and therefore sensation was the basis, in a direct way, of pure intellect. We find the relationship between sensation and intellect in Aquinas to be more harmonious and less conflicted than in Plato. Sensation is more of a challenge toward knowledge than a curse; it has grace in it. Sensation is considered the body of intellect, a necessity for it; and even if it involves suffering, this suffering is a necessary station for the fulfillment of intellect. In brief, sensation is an embodiment of grace, in which status sensation is given a very special position, in the center, or in the very heart of the intellect. In Aquinas sensation moves from the periphery of the intellect to the very center of the circle of knowledge. Also, for Descartes, as we shall see later, sensation is a pure celebration of the union of the soul and the body, which the intellect somehow cannot fathom unless it empties itself at its altar. Although Descartes, in his dualistic doctrine, allows and even emphasizes the distinction between the soul and the body, he admits the distinction to be a ruse of style rather than a metaphysical position. For Descartes, too, final reality is postulated in the notion of sensation as personal experience.

3. Descartes' Ruse and Cartesian Dualism

Descartes represents the beginning of modern philosophy and, in a way, also of modern psychology because of the wide range of reactions his work generated. Earlier Plato had led to the metaphysical distinction between the intellect and the sensory, between the spiritual and the material, from which he had drawn his final conclusion that only the world of idea was real, the material world being only its shadow or copy. Plato idealistically argued that if man solely relies on senses, he cannot reach the world of ideas. According to Plato, ideas are the sole object of human knowledge, and the unfortunate union of the soul with the body erases the knowledge of ideas in the soul which existed prior to its earthly existence in the body. Interestingly, Descartes' position not only represented, in some way, a continuation of Plato's idealistic stance but also the realism of Aristotle. In his famous dualistic doctrine, Descartes stated that man is a soul and a body, the soul being spiritual and the body being material, and acted like a machine. He, however, emphasized that the two spheres of knowledge could and should be studied separately. But as has been said by many, Descartes himself was not really a dualist (Spicker, 1970) but rather a tactical peacemaker between extreme positions. He wanted the philosophy of ideas and the science of matter to flourish independently and made way for a very useful approach which helped develop modern knowledge. He, no doubt, succeeded in this practical achievement, which unleashed new investigation in many areas of thought and science, but as to peace he only managed to create an unstable and uneasy one. His doctrine has been a fertile ground for unleashing of new wars in the sphere of human inquiry. One doubts if Descartes had really intended anything beyond the uneasy peace he created, for there was another deeper side to Descartes, that of a strategist. His dualism was really a ploy to create a battling space for human knowledge at various levels of inquiry.

Cartesian dualism. This famous doctrine of strict dualism was named after Descartes (Latin name: Cartesius). At the same time many followers of Descartes believe, as was said earlier, that Descartes was not a "Cartesian" (La Mettrie, 1969). In his private notebook the enigmatic Descartes remarked, "I come forward in a mask," emphasizing that he only came up with a "clever trick, a ruse of style," against the theologians who held a tight control on knowledge. For himself, Descartes never hoped to take the mask more seriously than he had in order to render an important service to knowledge. In a prolonged correspondence with Princess Elizabeth of Bohemia (1616-1680), he revealed his real face, saying that he did not maintain the truth of this "professed" dualism, although he continued to emphasize that "my main object was to prove the distinction of soul and body." In the letters he revealed his true position in the desire not to push the distinction too far, and that the discussion concerning union of soul and body was admissible, although this union could not be understood except in an obscure way by pure intellect or even when the intellect is aided by imagination. Finally he asserted that the union can be understood very clearly by means of the senses. Not in meditation or study but in the course of ordinary life are we able to perceive the union of soul and body. As the correspondence proceeded on, Descartes reminded the Princess that she could arrive at a viewing of this unity as a knowledge within herself. He explained to her that she could form such a conception (of union) by experiencing it in her own person. Descartes concluded in her own person. Descartes concluded in these words to the Princess: "everybody always (has the awareness of the union of soul and body) in himself without doing philosophy..." Whenever Descartes spoke of this concept of unity, he frequently referred to it as his "nature." In his *Meditation Six* he clearly used phrases like "dictates of nature," "necessary to my nature," "nature seemed to incline me," "nature teaches me," "being taught by nature," and so forth.

Behavioristic monism. It is, however, very strange to observe that in later formulations of the Cartesian doctrine, the position has been turned into a metaphysical doctrine. It is not the practical dualistic side of Descartes' thought but the more belligerent side which now permeates both philosophical and scientific thought. In an uncanny way, however, the doctrine has kept afloat even materialistic and idealistic monism. This destiny applies to psychologists as well as philosophers of the past and present, since dissatisfaction with the Cartesian doctrine only implies that for the very purpose of discussion monistic thought is forced to presuppose dualism. At the practical given level, monism presupposes what it combats at the theoretical level, thus keeping its enemy alive. Cartesian dualism is thus a tricky doctrine which must be dealt with in one way or another in a satisfactory or unsatisfactory way and that is where its perpetual strength lies. As already hinted, Descartes' practical dualism led in a paradoxical way to some monistic trends in the form of extreme idealism (Berkeley, 1685-1753 and Hegel, 1770-1831) as well as extreme materialism, of which modern behaviorism is an example. J. B. Watson tried to design a psychology not only without the soul, he even removed mind and consciousness from psychology, considering them substitute terms for the soul. But behavioristic monism is obviously not Cartesian although it stems from it, since it is not empirical but metaphysical in its attitude toward mental phenomena. Behaviorism avoids the basic question concerning distinctions between mind and body by denying that one class of events exists at all. The problem is that Cartesianism had only recommended a practical dualism and did not state an idealistic doctrine.

II. LOCKE'S THEORY OF SENSATION AND REFLECTION

It was no one other than Locke himself who effectively picked up an aspect of Descartes' nondualistic philosophy and turned it into an empiricism which was at the same time critical of Descartes' theory of innate ideas. The essence and direction of Locke's equally clever maneuver can be easily misunderstood and turned into a pseudo-metaphysical doctrine, when we ignore the context in which this very fruitful position was developed.

In Descartes, sensation is shown to be the very union of soul and matter and a state of perfect knowledge, the apprehension of which does not require doing of philosophy but requires immediate personal experi ence. Pure intellect becomes only perfect and complete in the sensory experience without which it descends into obscurantism, a thesis which is ultimately nearer to Thomas Aquinas and opposite to Plato. Thus, finally we come to decipher that Descartes did not believe in the importance of innate ideas, rather in the "poetry" and "empiricism" of sensation. This reflects almost the same position as Locke's, but not completely so because Descartes did not make sensation the basis of scientific knowledge, he made it the basis of mystical experience. Locke "de-mysticalized" sensation and made a science out of it.

Locke rejected the notion of innate ideas and in its place hypothesized *reflection* to be the fountain of knowledge alongside the senses. Reflection, accordingly, provided operations beyond the primary and secondary qualities of the object (which come from the senses) and accounted for the ideas. In Locke's words, reflection is "perception of the operations of our own mind within us, and it is employed about the ideas it has got." Locke also opined that reflection is not used by children and comes later on with maturity, and most adults never attain it since their knowledge remains at a lower level. At this point in Lockean thought we, however, find ourselves returning to some kind of revival of theory of innate ideas clothed in the concept of maturity. Locke deserves this charge because he believes that reflection is used by very mature people only. No doubt, Locke is talking about scientific maturity and not philosophical maturity which takes him out of the realm of innate ideas into empiricism. But this marvelous feat of Locke is not without its problems and not without some blind spots which Locke tried to cover under a puristic vision of empiricism: a typically English view of a gentlemanly science dedicated to a mature viewing of things. All this is compressed in a single term in Locke, namely, reflection.

What is reflection in the Lockean sense really? If, by reflection, Locke means just a mature evaluation then we have to posit not one but two separate spheres of empirical knowledge: (1) concerning *mature* operations and (2) concerning *immature* operations. Even if children use immature evaluation, in some mysterious way they do profit from it also. Children obviously thrive on this "immature reflection" and because of it, not in spite of it, grow into adults. Thus, there is a whole area of empiricism under immature reflection which had been discarded by Locke in his pursuit of maturity. This area can be rightly called the various forms of reflection concerning sense data and ideas.

Obviously Locke failed to include children's "natural psychology" in his empirical science or to account for the authentic side of a child's mind. But the complications which Locke creates do not end here. When Locke states that some individuals never achieve reflection, it appears that by reflection he means not just adult reflection but scientific reflection which evolves from long training and concern for understanding of phenomena. Obviously, very few individuals become reflective scientists in the Lockean sense but they do become mature in some accepted popular sense of the word. Since Locke defines reflection as essentially scientific in spirit and mature in that way, one is forced to assume with Locke that most adults stay at the sensory level and operationalize their ideas in an immature way similar to what children do although there is an age difference. Such an analysis creates a host of further new problems for Locke's empiricism. One finally comes to the unavoidable conclusion that Locke's reflection which involves operations of ideas on sensory material is aimed at providing only "pure" and "higher" knowledge. In such a case which type of empirical operation would be directed to the study of non-reflective experience of children, average adults, and savages, remains unexplored.

1. Rudimentary Reflection Without the Pejorative

The problem with terms such as "immature reflection," "mature reflection" and so forth is that they are pejorative. The whole history of psychology is evidence in this direction since psychology has traditionally held on to the idea of reason and maturity as a normative value which must be applied to children as much as to adults. In a way the rudimentary reflections of children are authentic biological tools necessary to that level of development. The way children reflect on their experience, if we decide to extend the use of that term, is different from the manner in which adults reflect on their experience. If, on the other hand, we decide that children do not reflect, we are guilty of refusing to recognize that children do think and build hypotheses, and that an important link of biological continuity appears there. We need a science of rudimentary reflection which does not involve a pejorative view of what rudimentary means. "Rudimentary" should also mean "perfect" and "effective" at its own level of operations, involving those first manifestations.

We next find another complexity pertaining to the idea of adult reflection. Following Locke, we assume that most adults indulge in "immature reflection" but, according to the prevailing popular notion, may be considered quite "normal." Thus, by making a distinction between "mature" and "immature" at the "adult" level, we also set the important task of distinguishing an empirical science which studies this problematic reflection and the rules that govern that reflection. As we decipher this we also conceptualize that "immature reflection in children" is different from "immature reflection in adults." Psychology, and especially traditional psychoanalysis, has been guilty of confusing the two in their theoretical analysis. This resulted in their application of immature adult thinking to children and children's immature thinking to some form of adult hang-ups. Typically, a 3-year-old child was considered a 15-year-old adult idiot or vice versa. This contributed toward suppressing the criteria of distinctions pertaining to how children and adults form ideas.

We find that children's and adults' reflections, whether mature or immature, have different attributes. For instance, children form ideas by using trial-and-error reflection and play reflection. They also perform what may be called "mirror reflection" in the sense that they repeat images that adults have handed down to them. Thus, they invent their own images as well as take in ideas and images in the form they are given by adults and use them in their trial-and-error and play reflections. One imagines that if a person tells a child that a ghost or a spirit lives under some tree or in a stream, the child will mirror what the adult told him and say that a ghost or a spirit does live under the tree or in the stream. However, the child may also decide to conduct on his own some trial-and-error or testing procedure concerning this handed-down reflection. The child may go under the tree or into the stream and investigate and even pretend that he has seen the ghost or the spirit. He may even decide to play with the ghost till the adult tells him further that the ghost is a bad one and he should not play with it. The child may, as a result, abandon trial-and-error and play with the spirit, and settle down in complete imitation of the command and stop going to the place where the ghost supposedly lived. Finally, the child becomes exactly like the adult who passed down the belief in the ghost, the idea that it is bad, as well as the commandment that it should not be approached. When this child grows into an adult, he will probably do the same to his own child.

2. Ritual Reflection

One can take a second careful look at the adult who was reflecting on his own thinking and imaging mode to find out ways and means of imparting his belief system to the child in the most effective manner he could. This adult no doubt practices reflection, although this is only dedicated to a belief system. The adult happens to be powerful, and can control, cajole and threaten the child to behave the way he wants the child to behave. The society thus develops a whole sophisticated portfolio of ritualized behavior which is aimed solely at imparting the belief system it believes to be true. This portfolio of imagery operations is handed down from generation to generation and becomes a form of cultural or social reflection which can be called "ritual reflection," since it is composed of emotively learned repetitious behavior. This reflection has the support of the whole mob behavior over generations, and as to its truth or falsity, it can neither be tested nor questioned in the social context. If any impulse toward testing or questioning arises, it is warded off by cajoling, threats and eventual control in exactly the same way the adult had earlier cajoled, threatened and controlled the child concerning his behavior toward the imagined ghost.

Magic. Magic deserves to be briefly discussed here, for the power it is said to have on the mind. The theory of magic has obviously moved away from its Victorian position to a more modern attitude toward it in writers such as Levi-Strauss (1964). An interesting case described and discussed by Sudhir Kakar (1982), an Indian psychiatrist reporting on shamans, will elucidate our own position in this respect. A young Tibetan monk had recently suffered an attack by a spirit called "Lu" (wind). On the day of the attack, the monk had gone for a local celebration with other monks. While returning, the other younger monks playfully turned the journey into a competition and started running. The patient ran faster, and with great abandon, and got ahead of everyone, but suddenly was found lying unconscious on the path. On recovery, he found his body badly bruised and bleeding at the palms and elbows. He was taken by his friends to a waterfall in a nearby stream where he washed his wounds. Two "hippies" were also washing their soiled clothes and dirty dishes in the same stream. After returning, some medicine was applied to his cuts and bruises, but it did not help. His face swelled up to a point that he could barely see. At this point the head lama of the monastery was approached, who gave the verdict that the illness was a case of spirit attack. The events were thoroughly analyzed and tracked down to the patient washing his wounds at the stream. It was locally believed that a spirit called "Lu" lived in the stream. The patient had supposedly angered the spirit by pouring all that filth in the stream, especially since spirits are very sensitive about cleanliness (blood is considered unclean in Tibet). The lama confirmed that the possessing spirit was "Lu." The patient took some worship offerings and offered them to "Lu" at the stream. Immediately after offering the gifts, apologizing and making amends, the swelling on the monk's face disappeared. Cross-questioned by Kakar as to why the spirit did not attack the hippies who were similarly dirtying the stream, the lama explained that the spirit in the stream did not attack the hippies because the hippies did not believe in the spirit, for which reason the spirit did not care what the hippies did. Kakar tried to explain the cure in terms of Freudian psychoanalytic theory, that the competition with the older monks stirred up forbidden feelings of unconscious hostility against the father figures of the monastary — an oedipal explanation. It is no doubt possible to translate the description of a fact from one symbolic system to another, but like all translations the outcome is always poor. In fact, attack by "Lu" could have been equally explained in terms of Jungian psychology. However, what Kakar failed to recognize was that a person who is rooted in an experiential system can be cured only within the experiential system he knows. Any attempt to translate the monk's symptoms first in terms of a Western theory would have led to more complications before he could have responded within the new language system. The point remains that the monk was healed quicker through carrying out an operation within the same symbolic context as that in which the illness had arisen. The lama's reaction of contempt to Kakar's interpretation in terms of Freud is quite understandable because, in the lama's view, Kakar was similarly polluting the pure stream of Tibetan communications where also the spirit "Lu" lived. Following the interpretation the lama reportedly dismissed Kakar without a response, like the spirit "Lu" had dismissed the hippies. Since Kakar didn't believe in "Lu," nothing could be done about it.

It is imperative to understand that it is the indigenous motifs which are both the context of the disease as well as its cure within a belief system, and that translation of the belief system into another symbolic system makes the belief system inaccessible for therapeutic operations. Treatment of a magical motif is possible only through magic. One can combat a particular form of magic only through magical reflection which belongs to it and not through a rationalistic analysis.

3. Mythological Reflection

We can add yet another venerable item to the list of reflections, one which has earned the respect of civilizations over the ages, namely, mythological reflection. Under this reflection can be listed all the religious and mythological reflections which the world has employed from time immemorial. Mythological reflection is the larger emotive context for ritual reflection, the latter being more narrowly sculptured as a nuts-and-bolts operation, dedicated to forbidding or cajoling people concerning certain goals or behaviors which may or may not be validated by or directly connected with the larger mythic context.

There has been a tendency in the empirical theory to judge myths as true or false, or merely as immature forms of reflection. In the light of the previous analysis, one can see that empirical theory commits the same serious tactical mistake which it committed in monistic behaviorism. By dissociating itself from myth operations and leaving the field to "judges of truth," who either simply exploit or condemn myths for political purposes, empiricism abandoned its responsibility to describe the myth operations societies universally pursue. Myths make people vulnerable to riots or bouts of hypnotic peace. Scientific methods of prediction and control can be employed for monitoring myths as a vehicle of personal reflection as well as analysis of social engineering process.

The ritual reflection, which comprises a major complementary portion of reflections attached to myths, is a more direct instrument of control and prediction and serves as a context for scientific scrutiny of, first, the rituals themselves, and then ultimately the myths. This takes us to our next category, namely, scientific reflection and its role in consciousness.

4. Scientific Reflection

Scientific reflection and ritual reflection repel each other in the same way as the adult in our example had stopped the inquisitive child from testing the ghost under the tree. If the scientist in the ritualized society insists on testing, he is usually dragged away from the so-called "danger" by the so-called more "mature" adults, or the "architects of the society, as happened in the famous trial of Socrates in Greece. On the other hand, in scientific reflection we find a paradox in that it is a category of reflection which is far more powerful than children's and adults' reflection, yet this reflection is at the same time most helplessly bogged down in the prevailing belief systems, which makes its truth-seeking operations most paradoxical. Practically speaking, scientific reflection has been the most enlightening as well as most repressive toward the savage, which humanity has known over the ages. Near the end of the Middle Ages, this reflection itself ferociously battled the ignorance; yet it lived to give birth to behaviorism. But, let the reader be aware that the aim here is not to separate out good from bad in the history of this struggle but to describe the interplay of idea systems and the complex operations which are involved in them. The point here is not to describe scientific reflection and myth reflection as necessarily true or false, but to characterize and define the scope of empiricism concerning these reflections which have been fountainheads of human behavior and the battleground for supremacy and control.

5. Political Reflection

Politics all along in human history has used "tested" operations involving rituals and myths to control and govern people. In modern politics the use of empirical exploitation of these operations concerning government is most sophisticated. The basic principle is not to attempt to educate people but to operationalize events based on prevailing ritualized beliefs. The method follows the path of least resistance needed for implementation of control. Education of the masses is not only considered time consuming but undesirable since in the educated groups the spontaneous basis for the operations is lost. The problem is even more complicated by the fact that myth and legend are part of the human heritage and without this much of human life would be empty. Thus, the issue is not whether a certain belief is correct or false; it is considered best to leave things as they are and formulate operations on the basis of ritualistic forms that exist even if they happen to be relics and fabrications. In this endeavor misinformation as well as disinformation become part of weaponry for control. This principle is extensively exploited by Madison Avenue in their ads with a flourish which mixes imagination with deceit. Marlboro cigarette ads show a very beautiful clear sky and a pristine blue lake in the background, while in the foreground, a cowboy on a horse is puffing away at a cigarette. Purity and pollutants are put alongside each other in the new methods of selling (Ahsen, 1984a). The unwary person who looks at the modern associative techniques accepts the conditioning offered through the ad. The next time when he picks up a cigarette and smokes he has no sense of guilt whatever; in fact, he feels that he is honoring the blue sky and the lake. A reader was shocked to find that *Psychology Today*, which is published by the American Psychological Association, accepts liquor and cigarette ads. He queried if the *American Journal of Psychiatry* will follow suit and begin to publish ads for liquor and tobacco. A psychologist who heard these remarks defended the APA by saying that the psychiatrists themselves are drug pushers, missing the point.

The lesson from this discussion is that unless you do consider the focus at the level of empiricism and operations, you have lost the understanding of the modern society. We need to know first how children or childlike adults, social theorists, politicians, and scientists operationalize. After Locke, the course of empiricism abandoned this side of the fertile ground of empiricism and instead searched for abstract truth, missing the point that scientific reflection as a historical process is kind of an overall political reflection. The scientists have always been interactive with a part of the political process and much of the scientific research is even frankly guided by political goals which have contributed to the character of scientific information which has evolved today.

What one would like to underline is that the complex political reflection can be empiricized by the scientist and described in the way it really functions to evolve a clear understanding of dynamics in society as a whole and the working of individual minds in that context. Although apparently separate from each other, all these reflections involve each other. The reflections which involve various complex operations with their sense data are truly baffling. The only way you can make a clear distinction is at the level of Locke's idea of maturity, the maturity of a reflective vision or maturity of a reflective operation. An operation can be very mature in the sense of being very effective, but may also be without a vision of truth. We slide back again into Locke's problematic position and the responsibility it implies concerning the term "maturity," whether we apply it to the child's mind or the adult's mind. The use of the terms "reflection" and "maturity" in Allport will be helpful here, for it concerns imagery in a more direct way.

6. Allport, Eidetic and Locke's Reflection

I am presently thinking of Allport's (1924) discussion of the eidetic in which he accepts the existence of the eidetic image and emphasizes its usefulness among children in the   words:

The EI seems to have essentially the same purpose in the mental development of the child as does the repetition of a stimulus situation. It permits the concrete "sensory" aspects of the surrounding world to penetrate thoroughly into his mind...Such pseudo-sensory experience enables him to 'study out' in his own way and in his own time the various possibilities for response contained within the stimulus situation. His reaction when the situation is first presented is often incomplete, the presence of adults, or the lack of time, preventing him from becoming thoroughly acquainted with its properties. A period of reflection is necessary, during which he may experiment in various ways with his image, varying his behaviour to conform sometimes to one and sometimes to another aspect of the situation, gradually gaining a comprehension of the full meaning of the whole, and building up the attitude which is to determine his future response to the same or to analogous situations....While the persistence of the literal images enables him to perfect his responses to a situation when the original stimulus has ceased to operate, the alterations in the images enable him to elaborate, within limits, the stimulus situation, and so to provide himself with data for a certain versatility in behaviour (p. 118).

In the above quote one notes the occurrence of the word "reflection" where Allport considers that the child does need reflection and does practice it, and that the eidetic image provides him the opportunity to do so. However, as Allport (1924) continues on, he makes rather a strange statement:

In later life there is no need for detailed images. In any case eidetic imagery is an anomaly in adult life. Its true function is performed only in the earlier years of mental development, when by preserving and elaborating sensory data it enhances the meaning of the stimulus situation for the child and enables him to perfect his adaptive responses (p. 119).

Here Allport displays a striking lack concerning clinical facts. It is common clinical knowledge that many emotional issues of a person's life remain unresolved and require replay and re-evaluation. I have elsewhere called this "Allport's Bungle" (Ahsen, 1977b). One wonders if Allport naively believed that during adolescence and later on people really resolve their problems and conflicts and do not need any sensory replay of experience to reflect upon. It is universal knowledge that conflict in human beings continues on from childhood into adulthood and later on into old age. In fact, adults appear to be more conflicted than children and as such more in need of enactment of this special type of sensory imagery revealed in the eidetic. Once we accept the role of eidetic imagery in the replay of a sensory stimulus for children, we cannot get rid of it as a necessary function in later life, because the need for replay of a sensory stimulus never really ceases at any stage of life. Here lies the fundamental importance of the eidetic in the empirical theory.

The eidetic places a different accent on the approach to and status of memory in psychology, posing the question whether memory is central or peripheral in the study of developmental processes. For a child it would be difficult to reflect on pure memory. It is more adult to deal with memory and to reflect on it in order to make something out of it. The child has almost no capacity to treat memory as memory because he still finds himself attached to the previous experience in a manner that he cannot extricate himself from its ongoing nature to reflect upon it. The capacity for a special type of reflection is preserved in the eidetic image in that it is not yet memory per se but an ongoing experience. The struggle in the first perception which continues on in the eidetic is reflected upon in a more involved and dramatic fashion. If we look at the nature of a child's reflection during eidetic imagery, we find that the reflection in the child is very profoundly empirical, relating to and examining sensation, and the operations appearing in it are of the ongoing type rather than of the sealed type as in intellectual thinking. One would suppose that reflection always begins that way and with the passage of time becomes more constricted, more fossilized and more like memory, a mere remembrance of the shell rather than a slice of life. However, the original reflection never loses the quality of being a part of the inherent operations of the mind, that is, its real empirical core. Thus, in a deep philosophical sense child's reflection is the real empirical model in which the object and subject are closely united yet sufficiently free to play upon each other in a more experimental way.

While Allport wonders about the anomaly of eidetic imagery existing in adults, since adults are supposedly mature and do not need the perpetuation of this kind of imagery to reflect upon, Locke believes in the opposite, that very few people ever become mature enough and achieve the ability to properly reflect. The kind of reflection which Allport believes exists in children, Locke does not accept as reflection at all, and he even doubts the existence of reflection proper in the adult. At the same time, Locke thinks that reflection is "the perception of the operations of our own mind within us, and it is employed about the ideas that it has got." If reflection is really a part of the mental apparatus, we require a restatement of the empirical method concerning various types of reflection operating at various levels.

III. EMPIRICISM, IMAGERY, MYTH AND METAPHOR

At this point, one needs to make a bold statement about the empirical method concerning sense data. Empiricism cannot postpone an open-ended treatment of imagery operations at both the personal and the social level. We need to involve ideas of animism and mythology as much as ideas of technology in order to understand the nature of operations in the modern world. A mental object, whether social, natural or manufactured, is a simultaneous combination of all these aspects, and is real only in that sense (Ahsen, 1984a, pp. 225-248; 1984b, pp. 164-185; Count-van Manen, 1985; Mead, 1934, 1938). The baffling mix of mechanics, myth, metaphor and politics must be described through a scientific methodology which does not merely distinguish between abstract truth and operation but describes the operations as they exist and are carried out, or can be carried out in the historical process. Since misinterpretation is currently a part of the whole process concerning discovery of truth, we cannot hope to have a truth-seeking science but only an empirical science, which can be seen finally as truth seeking in the highest sense of the word, because it recognizes the role of misinterpretation in the process of such active knowledge (Ahsen, 1984c; Bloom, 1982). This is not a negative view of science but the only possible view of science at this point of development of human civilization. Whether we will be forever condemned to involve misinterpretation in our truth-seeking behavior is an open question, and history or annihilation of history alone can answer it. In this empirical view of imagery a picture is no longer a mere copy but a veritable power, a material process, a meaning, a symbol, a metaphor. The picture has split off from its origin and is making a new association. This picture is, in Bloom's literary criticism, the uncanny ephebe, struggling to disengage from the power of its awesome predecessor poet. The copy is asserting itself to become an original.

1. Dissociation and Neo-dissociation

The image splitting away from the source in order to form a new association is, in fact, the beginning point for the description of new materials. There is a widespread tangential theoretical conflict in the understanding of this area but the various viewpoints touch the same aspects from a variety of different angles. We witness such a splitting first in Descartes in the form of Cartesian dualism; a dissociative philosophical response to control by theological dogma. Descartes, as discussed earlier, admitted his avowed dualism to be a ruse of style rather than a genuine metaphysical doctrine he really believed in. He returned to experiential monism to express his faith in *sensation* clearly, based on distrust of the intellectual rational process which he said caused the splitting. But the fact of splitting in the human mind is deeper than this philosophical stance. Modern psychology has traced evidence toward splitting within the sensory field itself. It is not only the garrulous kingdom of the *word*, it is also the realm of *sensation* which is divided. The evidence for this has come from the study of hypnosis. The argument has been put forth in the form of neo-dissociationist theory proposed by Ernest Hilgard (1977). This theory is a decisive improvement over the early dissociation theory developed by Janet and his successors, and is said to have important implications within psychology as a whole (McKellar, 1979). Hilgard says that so many subsystems of specialized cognitive control structures go on at the same time that the role of a single central control process which is firmly in the saddle, as an effective executive and a monitoring function, appears to be doubtful. Hilgard (1975), however, concludes in a positive way, saying: "The theory asserts quite simply that among the cognitive control systems operative within the individual there are various degrees of mutual awareness in the normal waking state, and the dramatic role of hypnosis is to alter these relationships" (p. 9). He further adds, "I am careful not to insist on an absolute dissociation, but to assert that all dissociations are relative ones, with interactions that can be detected if care is taken to detect them " (p. 10).

Hilgard's work, and Naruse's experiments on hypnosis (Chapter 3) are important evidence on relative dissociation. Peter McKellar (1979), who is interested in dissociation in its extreme forms, has attempted to connect the very process of theory formulation in science to dissociation, showing how the field of inquiry has been enriched through translation of terminology (see Chapter 2). To McKellar, the healing process is actually this process of terminological translation, a willingness to say the same thing in words other than the person is accustomed to through sheer habit, fanaticism or mental blindness. He links the usefulness of free association in therapy to its dissociative character. He points out the use of metaphor in literature as also based on dissociation and derived from terminological translation. He tracks down various clashing views in the imagery debate in psychology as essentially using metaphors and reducible, in Dennett's words (1981), to "iconophiles" and "iconophobes," reminding us that the views of all the discussants were essentially analogical and involved some kind of ultimate relationship with imagery. Through pointing out analogy as the main structuring component in the mind, McKellar manages to reduce everything worthwhile to imagery. Although McKellar reserves a venerable central role for imagery, in the process he fails to fulfill his promise to base his theory on extreme dissociation. One can detect in McKellar's a final yielding gesture to Hilgard's neo-dissociation theory. In this author's view, Hilgard's theory of neo-dissociation involves the positive edge that no dissociation is possible without some concomitant readjustment of forces in terms of new associations. Said in an opposite way from the side of association, the statement would read: "No new association is possible without some concomitant dissociation." This ultimately means that some dissociation is fundamental to the very concept of association and that change or movement involves a breakdown of the existing balance of associations. No new image can emerge except out of the ashes of the old image. The process is somewhat like the birth of the mythical phoenix.

2. Metaphor: Resemblance and Deviance

Thus, the image splits from its source in order to render a new association possible, and this theme is reflected in literary criticism as well as psychology and covers a vast territory of discussion under the term "metaphor."

Metaphor is truly a complex area of theory. Dictionaries proverbially leave to the experts the headache of elucidating what they only put together in half-baked descriptions for the beginners. Thus, the *Oxford English Dictionary* defines metaphor as "application of a name or descriptive term to an object to which it is not literally applicable." Compare this with a pronouncement such as by John Bunyan, "The prophets used much by metaphor to set forth truth." Curiously, in respect to metaphor, the Oxford dictionary suggests possibility of error while Bunyan emphasizes truth, a vast difference indeed. Metaphor, therefore, is not a simple matter either of making "something unfamiliar clearer by comparing it with something familiar" (Buckler & Sklare,

1966, p. 182). By drawing mere analogy one achieves very little, and metaphor is not a mere analogy; it does more than merely compare or just give a different name for the same thing.

The use of analogy, simile and metaphor is, no doubt, widely common in thinking. However, scientists have always considered data from analogy a poor form of evidence, in fact, no evidence at all. Logicians likewise have considered argument through analogy an unacceptable form of reasoning. According to them analogy does not establish a valid form of argument nor does it provide a causal relationship between two things. In a likewise fashion the simile, in which two things are equated word for word through an explicit emphasis on similarity, has been considered a poor predicative description of a subject. In the case of metonymy substitution of one word for another takes place, when the objects each refers to are habitually associated. In a metaphor the resemblance is not a case of metonymy characterized by a word-for-word substitution but, as a rule, by the sentence as a whole. If the vehicle of a metaphor happens to be nonverbal colors and shapes as in a painting, the whole canvas carries the new articulation. It is not the separately recognizable buildings and shapes but their relationships as well which, as a unit, comprise the metaphor. According to the literary critic, Paul Ricoeur (1979), the image centrally occurs in a metaphor, but the image itself is not the metaphor, it only provides "figurability to the message." He makes this statement in the classic tradition of rhetoric, saying that "the tropes make discourse appear," that the figuration provides a quasi-body to the communication and sets before the eyes the *sense* that it displays. It moves *sense* to another *sense*. Briefly stated, any theory of metaphor has to deal with two basic processes which appear in it simultaneously: resemblance and deviance. The function of resemblance in a simile stops at resemblance and does not go further, but in a metaphor the work of resemblance goes beyond resemblance into deviance, which provides basis to the metaphor. Thus, the function of resemblance in metaphor does not aim at creating a mere copy but a vividness which in turn helps to provide movement to a discourse. The message thus goes beyond similarity.

According to Aristotle, *lexis* (which is diction, style, eloquence, and wherein metaphor appears as one of the important figures) makes discourse or *logos* appear as such and such. Aristotle argued that a good metaphor relies on the contemplation of similarities, and his argument is based on the picture function of a metaphor in the process of creating a meaning. Some have translated Aristotle's phrase "to contemplate similarities" as "to have an insight into likeness." Thus, in metaphor discourse assumes a body displaying a form, which is imparting of further "sight," that is, "insight" to the message over and above the similarity. The picture itself has been taken beyond similarity into being itself, by remaining not a copy but becoming an original. The work of resemblance in a metaphor is to take the copy, and with a view to persuade and to please, to turn it into an articulated new statement. This is achieved through the intervening process of deviance, which goes beyond the concern for a copy since all the predictive structures are not found in it. Resemblance means mere substitution while metaphor means interaction. The theory of literary metaphor thus has Hilgard's dissociative component which accounts for a new association. Elsewhere I have discussed (Ahsen, 1984a) the limitation of the copy theory of imagery in Cautela (1977) and others, such as Paivio (1976) whose Dual Code theory similarly follows the principle of substitution, that is, of a verbal code for an image. The limitation of a theory which posits a simultaneous parallel coding of an event in image and word should be obvious since the theory does not use a dissociative model to account for association-dissociation spectrum.

In metaphor, there is no substitution but an interaction between subject and predicate. Since metaphor consists of some deviance, the fact compels the subject to be described in a new way; it is not an erroneous denomination but a deviant prediction. Jean Cohen (1966) calls this deviance the semantic impertinence, suggesting that ordinary, commonplace relevance or, in other words, the code of pertinence which rules and denominates the ascription of predicates in routine day-to-day use is violated. The new pertinence secures the production of a lexical deviance, which involves a paradigmatic deviance, spoken about by classical rhetoricians but not clearly understood till modern rhetorics came on the scene. According to Ricoeur, classical rhetoric closely concentrated on the level of the word and not on the level of the whole sentence, overlooking the fact that the semantic twist occurs at the level of the whole utterance. The rule of resemblance is confirmed and broken at the same time. Semantic impertinence derives its power from this acceptable deviance in order to create semantic innovation.

Stasis is not the rule in nature but movement is. When an area in consciousness or existence becomes overloaded with tension, it moves in order to create a new combination in which the original structure or the basis of existence dissociates, although not completely. The primordial hologram overflows at every point, and the spirit of the hologram tends to recreate itself. This explains the importance of deviance as a primary function in the creation of metaphor. Metaphor springs from a feeling of pressure. One of the tendencies in the mind is toward closure, egocentricity and repetitive logic. The metaphor is born in response to this sense of poetic incompletion. In simile, one says that something is *like* another thing but in the metaphor one says that something *is* another thing, and the statement of dissimilarity involves a whole new explosive dissociation in which freshly formed associations appear.

The struggle for creation of new interpretations has led to many views and doctrines of the extreme type such as Harold Bloom's doctrine of misreading and misinterpretation and the famous school of surrealism. Bloom's idea of wrestling by the ephebe with his awesome predecessor, and the surrealists' games like *Cadavre exquis* in both pictorial and verbal forms (Matthews, 1982) are examples of attitudes toward the challenge of discovery. It is a kind of ludic activity in which the participant engages with real excitement. The process is aimed at enriching cognition by rising handsomely to the occasion. Surrealists have elevated their serious, but at the same time amusing, pastime into an imperious theory which offers a second fresh look at Charcot and Janet's *Automatisme Psychologique.* In brief, they hold that the delirious system which is reflected in hysteria is more or less an irreducible state which is characterized by subversion of the relationships set up between the subject and the external mental world, and in this sense is not a pathological phenomenon but a supreme means of expression. This view of the surrealist theorist Andre Breton (1924) is extremely close to a similar view held by R.D. Laing on schizohrenia (1965). To highlight their manifesto on the human mind, surrealists have consistently held a yearly celebration of hysteria in which they get together to honor Charcot's nonclinical interest in hysteria and go all out to condemn Janet's denigration of the *delirium.* They draw attention to what they call the increasingly abusive denunciation of what, since Bleuer, has been called "autism," since it allows the clinician to conveniently denounce as pathological anything in man that is not pure and simple adaptation to external conditions. The surrealists demand what they call "rights of imagination" (Breton, 1924). One of the many weapons of surrealist "revolution" is automatic writing which is found in the arsenals of hysteria as well as hypnosis, the same weapon which led Ernest Hilgard to discover his *Hidden Observer* who, in a wily but judicious way, commented on the other part of the executive ego which was under external control in the hypnosis setting (Hilgard, 1977). The essential differences between the popular clinical and surrealistic view of imagery is best describable in terms of Dali's paradoxical statement that the only difference between himself and the madman is that he, Dali, is not mad.

Dali is referring to a view of imagery in which an aspect of the normal in imagery is described as one similar to a madman's without it being in any way clinical. The idea is not that there is no distinction between the clinical and the surrealistic but that the distinction has been marred and blurred by the clinical viewpoint which omitted to describe the surrealist in normal consciousness as normal. The norm in the clinical dogma is the madness by which it has managed to ignore the healthy aspects of the clinical phenomena. Hypnosis researchers similarly complain about a popularistic negative view of hypnosis.

3. Copy, Metaphor and Split Reference

Whether image is a copy of the outside reality or is aimed at destroying the outside fixed reality is an important question. It reminds one of Hamlet's words in Shakespeare's play, "To be or not to be; that is the question!" The nature of the issue under question here appears in epistemological analysis of imagery, hypnosis research as well as rhetorics concerning poetic discourse. The person who experiences the compulsion of making an assertive statement such as A is A is also simultaneously experiencing doubt concerning it, which means A is not A (Ahsen, 1968). The phenomenon of *negative hallucination* first studied by McDougall (1926) and later by Orne (1962) deals with such a paradox: A subject who is hypnotized not to see something in a room, such as a chair, reports that he does not see the chair but describes the space where the chair is as "somewhat more empty than the rest of the empty space." Orne's conclusion, like McDougall's, was that the hypnotized subjects were able to perceive and not perceive at the same time.

The basic argument concerning metaphor proceeds on the ground that it involves a kind of stereoscopic approach which brings together two otherwise distant realities, putting them together into a conjoining vision. The two terms are separate, yet they are brought together through what Roman Jacobson (1962) called "split reference," and that establishes the nature of the metaphor. We need to be somewhat more precise here because this definition of metaphor, though instructive, can create a poetic and an epistemological blind spot concerning the single image phenomenon itself. The question is: Can we detect a metaphorical process within the single image process itself? If we can, then the metaphor the poets talk about is, at best, an explicit metaphor while, at the same time, an implicit metaphorical process can be detected within the single image which is perhaps the real original basis for the existence of explicit metaphor. We need to track dissociation in the single image itself.

Jacobson refers to the fact that the referential function does not totally disappear in the poetic discourse, but it is profoundly altered by the metaphorical message itself. The supremacy of the poetic function treats the referential function in a split paradoxical way, which Ricoeur calls something cogent and similar to the preamble to fairy tales: "It was and it was not." The poetic discourse uses a complex strategy whereas a straightforward referential treatment uses a simple one. The discourse, as poets have repeatedly pointed out, suspends the ordinary reference which the descriptive language is habitually attached to. Therefore, on the suspension or ruins of the first-order-reference is built the second-order-reference which speaks as much authoritatively of reality as did the first-order-reference. The ordinary reference has been replaced by an extraordinary or primordial reference; the objective has been replaced by pre-objective. Ontologically, as Hilgard would agree, these various orders of existence have their own importance and relevancy, and re-rendering of the objective level in terms of the pre-objective level remains clinically as well as poetically an important operation. The copy, the primordial function, the newness are interconnected.

4. Image as Mutant

In the theory of poetic language there has been a discussion of the difference poetic language holds with referential language in which an image stands as sign for an object. Jacobson distinguishes between the poetic function and the reference function. In the poetic function the language redirects to itself (implying a mutation in the use of language according to the structuralist group), while in the reference function the ordinary language directs itself to the outside, to the reality. What is an image in the mind then? It is not a sign for something, it is in itself something. It is not referring to outside. Reference to itself makes image a mutant. The distinction is not unimportant and has application in the clinic. For instance, the morbid image of a trauma no longer refers to a reality, because that original reality is no longer there; it has passed away. Still, the image repeats itself with a characteristic compulsion which is damaging to the patient, and the patient does not know how to stop it in his mind. The ongoing traumatic image is not a copy (the original event has stopped), it is a mutant, a thing in itself; it is not just a sign, it is an event. Most neurotic problems and their images do not point to a reality; they are an inward view of things in response to a reality which once existed, no longer exists and is believed to be still ongoing. In fact, the idea that the neurotic image is a reference image is a real mistake, and is cause of the problematic way the patient tends not to see the original event. No doubt, sciences need referential language, but of all the sciences psychology is the only one which does not need a language of this sort. Image structure and its material is mutational in character. The treatment of the image in psychology should be at the mutational level.

The above argument can be made more explicit through a step-by-step analysis of the mental operation involved in the experience when someone wants to discover a direct correspondence between a copy and the original at the introspective level. A careful examination does not reveal the image to be a reproductive copy of something outside. We start by asking a person, "A," to see an image of his father, called here "Z," who exists outside in reality. "A" sees image "X" of his father "Z" as what he calls a "regular" image. After labeling image "X" as a copy of the real "Z," we next ask "A" to treat image "X" as a copy, and to see, instead, in his mind, the original of the copy "X," which is his father "Z" in reality. As a result, "A" reports seeing image "Y," in which his father appears not as a "regular" image but "somewhat more real." Next, we draw the subject's attention to the earlier image "X" and ask him to report how it looks. Somehow the instruction in which the person labeled the image "X" as a copy also changed image "X" into image "Xa," which now was reported as "somewhat faded, flattish and like a mere surface without depth." At the same time, we also find that the instruction to see the "original" gave us an image "Y," which was treated as "somewhat more real," an original with some touch of reality and a feeling that it existed outside. We notice, however, that in these transitions the real-real original outside in the reality, namely, "Z," never appeared in the mind except as an image, because it was, in fact, incapable of entering into the mind: it never behaved like a real person entering a real room. In the above experiment one never proved what one had proceeded to do, that is, show a direct parallel relationship between the copy and the original. Instead, we got many variations of imagery from the image of a copy to the image of an original, but we never reached the "original" in the mind. All images were copies or all images were originals; only a verbal reference to the "copy" made it faded and reference to the "original" made it livelier. In this sense the images "X," "Xa," and "Y" differed from each other in vividness, based on the instruction that "X" was or was not a copy. The differences showed that the person "A," in the beginning, was not treating "X" as a copy before it was labeled as such and then appeared faded as "Xa" under the instruction. To us, this suggests that the idea that the image in the mind is known by the person to be just a copy of something outside has no basis in imagery experience as such. The image of the object is treated in lieu of the object, and it is not its copy. In its new status of reality, the image offers different operations than reality. But more importantly, since the idea of the copy contains the idea of the original, "X" is a double image which contains "Y" in it in an implicit way. Thus, in the final analysis, every image as a copy of the outside entity is a double image, a stereoscopic vision, and it involves a split reference, in which sense it is a compressed metaphor.

A theoretician dealing with a unidimensional view of things is stuck in his own construction and does not see the underlying change in the term of reference and consequently the very nature of discourse concerning the image. We see that the split reference in a single image has far-reaching consequences for psychological theory. Cautela's copy theory in behaviorism is one example. This and other theories like Paivio's dual code theory do not appreciate that the verbal descriptions of an image do not enjoy a parallel relationship with the image and that existence of covert imagery in the split reference disrupts any possibility of strict parallelism. I have discussed the importance of such disruption for clinical application as well as literary discourse elsewhere under the "decoy model" (1983) and later on in *Trojan Horse* (1984a). When Odysseus told the Cyclops that his name was "No man" the expression disrupted the ordinary parallel relationship between image and word, creating a deadly consequence for the simple-minded Cyclops. The same thing happened in the famous image of the Trojan Horse in which the wooden horse enjoyed a split reference, as a "copy" of a horse, as an "original art piece," as a "devious contraption to deceive the Trojans," who saw the horse as a "toy" or a "trophy," all of these being split references. Thus, every image in the mind is a "copy" and an "original" at the same time, if we decide to keep in mind the reality demand as well as the fact that reality simply cannot be admitted into mind except on the ground rules which operate in the mental sphere. Mind is its own reality but it also dialogues with the outside reality. In this sense every mental image is implicitly a reconciled double image, which gives the single mental image the status of a metaphor under the rules.

Ricoeur in his theory extends the function of split reference to feeling and cognition, in order to explain the more mellow form of emotivity which appears in feeling and the semantic extension that is imparted to both image and feeling in the cognitive aspect. However, Ricoeur fails to see the split reference in a single image because classical rhetoric tradition has unfortunately discussed this phenomenon only at the level of the manifest metaphor, which uses two images existing at a distance from each other. One may be right in emphasizing the importance of split reference in understanding various levels of metaphor functioning, but we need also to explore the existence of the split reference within a single image. Since a copy works as a split reference, it obviously acts as a metaphor for the original, and as an original it represents a suspension of the same.

The other relevant quality which is mentioned in the work of imagination is suspension, which is similarly connected with the aforementioned paradox in the single image. According to Sartre, the act of imagining involves addressing oneself to what is not and this involves the capacity to project new possibilities. The single image is a suspension concerning what is not, namely, the original. Also it is not a weakened version of the original stimulus in the Humean sense, since there is no such thing as passive reception of the original stimulus; it is all the way active formulation. This is what Hebb (1968) clearly implied when he pointed out the role of the motor element in the formation of the image. The cell assemblies involved in the formation of the image were not merely those comprised of the presented sensory input, but the active motor input was an equally important component in the registering and formation of the image.

The analysis of imagery in literary theory somewhat lags behind the current analysis of imagery iñ the psychological field. Ricoeur proposes to discuss mental events which ride the borderline between the literary and the psychological, but the problem surfaces when he calls the single image merely reproductive. For instance, Ricoeur, while discussing image as fiction and the power of the symbolic system to remake reality, in Goodman's idiom (1968), comments, "But this reproductive and projective function of fiction can only be acknowledged if one sharply distinguishes it from the reproductive role of the so-called mental image which merely provides us with a representation of things already perceived. *Fiction* addresses itself to deeply rooted potentialities of reality to the extent that they are absent from the actualities with which we deal in everyday life under the mode of empirical control and manipulation. In that sense, fiction represents under a concrete mode the split structure of the reference pertaining to the metaphorical statement." In Hebb's description of the formation of the eidetic, the eidetic image is no doubt a fiction, but in any case there is no such thing as a purely reproductive role of the image in any form of imagery whatever. The first order fiction already exists in the image. Maybe what literature calls fiction per se is a second order fiction in which literature tries to dislodge mental functions from being locked into a frozen condition of the first order fiction where even the first order of imagination is being silenced and choked through a meaningless chatter of daily verbal non-discourse. The second order fiction can, no doubt, help us to rescue the first order fiction, in which case the distinction between the first and second order fiction would be operational rather than a real one. We need to recognize both levels, although images, emotions and thoughts at the first order level need to be elaborated first, without resorting to the second level operations. A host of operations can be conducted at the first order level. Psychotherapy and literature are full of examples in which a single image in its very first manifestation can become the bearer of the whole message which is spelled out not so much by conscious construction as by merely looking at or listening out. Full poems have come to poets suddenly, as have melodies to musicians, who heard them coming from outside, as if someone in the air was singing to them, and they wrote them down obediently. The eidetic theory of imagery emphasizes the importance of this type of first order creation in the psychological laboratory, therapeutic operations, or literature and art (Heidegger, 1967; Husserl, 1958).

The role of the single image in psychotherapy is in fact well established. When a patient is asked to see even one image, such as that of a parent, after the patient has been talking his head off, the change in the information flow can be remarkable. For instance, the patient may have been despairingly talking about an angry parent, but when the same parent is seen in the image, the portrait may be of a concerned parent rather than an angry parent. The concern in the image is about the patient's behavior which the patient does not see. As the patient begins to concentrate on this single image, a whole different spectrum of information begins to emerge in which the patient does find himself at fault, and does begin to understand the reversal in the image which came about later in life, that the patient began to find fault with the parent instead. This has been extensively discussed by this author elsewhere under the heading of Consciousness-Imagery Gap, or C-I-G (Ahsen, 1972). It is obvious that the mental condition which emerges during the verbal pressure can keep important images suppressed. Thus, words can at times be repressive to consciousess and stop the flow, especially of those images which do not fit into the lexical schema which the patient is accustomed to using. At this point one needs to see originality in the image and not merely in the words.

B. R. Bugelski (1982), who is a behaviorist, has moved closer to literature in criticizing Joseph Cautela (1977) for his overly narrow behavioristic definition of the image as a copy or as a mere word, a welcome direction in behaviorism, which has long been stuck in pseudo-metaphysics. While Bugelski underlines the centrality of the image, he is also critical of Paivio's Dual Code Model of imagery which posits a parallelism between the image and the verbal description, a theory which curiously treats image and word as a copy of each other, and disregards the semantic interactive relationship, lack of equivalence and presence of deviance, conditions necessary for the development of therapeutic maneuvers. The abstract substitutive status which Paivio gives to language is not supported by Bugelski's (1977) experiments on bilinguals which showed that words which referred to early childhood experiences belonged to the first language which was learned earlier and the words which referred to later childhood belonged to the second language which was learned later, and that the second language could not access the images of the first language. If language had been truly an effective abstract instrument, both languages would have successfully described the experiences at both the age levels. The emotive response to early childhood experiences was reachable only when the first emotive language was used to relate to those experiences. Bugelski's work provides a clear proof against the abstract value of language proposed by the proponents of the propositional school on imagery such as Zenon W. Pylyshyn (1973) and Albert Bandura (1977), who erroneously theorize that verbally-based codes are more concise and therefore hold more information than imaginal codes.

5. Drama

For a long period of time, psychological theory has worked in the service of mechanistic and diagnostic concepts. To describe or to intervene in psychological theory means to treat a feature without interaction, and to determine its significance with the idea that if there is something wrong with a feature, a cold surgical approach will cure the problem. However, a manifestation changes when one interacts with it. A neutral description does not generate true values nor do living consequences emerge from that. The same argument tells us that a diagnosis built out of sheer description would be a second generation child of an inactive view of reality. The interaction or drama theory tries to correct this blind spot in psychological theory hoping to replace neutral material in the clinic with interactive dramatic material. Obviously, drama in conjunction with imagery offers a methodology of interaction which takes the emphasis away from feelingless viewing. With a touch of invention the developmental scene is changed to bring out a new expression. The very idea of truth is different in dramatics: truth is not what is revealed but what can be revealed through what you are doing. It alerts the participant to the various possibilities inherent in action. Kenneth Burke's (1957) theory of dramatistics offers a psychological insight into this side of consciousness. To construct an image by acting upon it linguistically has been related to the problem-solving theory in imagery by Geir Kaufmann (1980). We note that the process involves an operation somewhat similiar to construction of dramatic processes. However, the fact that this theory has not so far involved the body (somatic response) in problem solving shows that problem-solving theory of language still has new developments to tussle with once the drama of the body processes are also introduced in the theoretical model. The somatic factor is important in imagery problem solving as much as in the psychotherapy process. It is well recognized that people generally report occurrence of positive body feeling as a result of semantic connection. In fact, an insight or solution of a problem may be preceded, accompanied or followed by an intense positive "S" factor, the somatic response. A negative somatic response (anxiety) can obstruct the emergence of insight or a solution, as can an uncontrolled somatic response such as in hyperactive learning disabled children. The role of somatics in problem-solving theory of imagery needs to be accounted for.

In an interesting way the theory of images in literature distinguishes between "wild" images, which occur as interruptions and distort or divert the mind from the message, and those which are called "bound" images and comprise the crux of the message. The "wild" images are considered extrinsic to the structure of sense and make the reader a dreamer rather than a reader. In the words of Sartre, a "wild" image has more fascination and delusion than production of sense. The clinical field, no doubt, has to account for the "wild" images which accompany a message, for the purpose of diagnosis, but when the clinical field fails to exploit the potential of the "bound" image it fails to address itself to the theory of effective communication. The theory of metaphor attempts to underline the importance of effective communication by pointing out the semantic nature of the "bound" images. But "wild" images appear along with "bound" images during composition of a sentence, especially during its utterance, that is, the reading or hearing stages. The process of verbal expression can also be like imagery, "wild" or "bound," and no verbal expression can be found which in one way or the other does not involve these two types of imagery in its figurations. What Hester (1967) said about poetic language is true about all language; what happens in imagery also happens in language. Language not only schematizes but also thickens the figuration process, making the imagined scene appear as a bound image. Good language blurs the distinction between verbal and nonverbal; it is not merely verbal in nature.

IV. MECHANICS OF IMAGERY

Thus, the central issues concerning the compelling power of imagery have not yet been fully resolved in psychology or literature. In literary criticism the distinction between feeling and emotion supplies us with the most persistent prejudice which has taken the form of a theory that starts with recognizing the importance of feeling and emotion in experience of the image, but soon proceeds to make a distinction between feeling and emotion, a distinction which, on the one hand, helps to make certain things clear but, on the other hand, also introduces a basic confusion that make mechanics of imagery operations less understandable.

1. The Triple Code Model: ISM

It must be pointed out that all modern theories of metaphor imply a tripartite model, including Paul Ricoeur's theory which implies a triple code reference in the very title of the 1979 essay, "Metaphorical Process as Cognition, Imagination and Feeling." These three elements may appear in different theories under different names or different sequential arrangements, but they imply the same tripartite relationship. For instance, the sequence discussed by Ricoeur in the body of the above-mentioned article changes from the one in the title and appears as imagination, feeling and cognition. Because Ricoeur proposed to discuss his theory of metaphor as a semantic information process, he decided on the sequence presented in the title. However, after clearing away the preliminary examination of semantic import he proceeded to discuss his theory the way he wanted it to read, that is, imagination, feeling and cognition. In this tripartite or Triple Code Model imagination or image (I) is the first component; feeling, emotion, or somatic response (S) is the second component; and cognition, semantic relevance or meaning (M) is the third component. This psychological theory involving the acronym ISM was first discussed by me in 1965 (and subsequently in 1968, 1972, 1977a, 1977b, 1984a-e). Although the ISM is not explicitly formulated in the classical theory of rhetoric or in later literary theories of imagination, it is nevertheless implied in all these theories as it is in all psychological theories.

The cause of the confusion appears to be that various theories tend to use the same terms for different areas of reference; for instance, in the tradition of rhetoric *sense* means *meaning,* while in psychology it means something akin to *sensation.* Thus, whereas in psychology the sense of an experience would be *sensation*, in rhetoric it means the semantic meaning, or the objective content as opposed to mental representation. This difference between the two uses extends to image and feeling as well. Whereas in general psychological usage the image may have a somatic or physiological feeling or an emotive tone accompanying it, the term *feeling* in the tradition of rhetoric may again mean *sense* or semantic meaning (Frege, 1978; Husserl, 1958; Ricoeur, 1979). Again, when the tradition of rhetoric speaks of *semantic* function of an imaginal process such as metaphor, the term is not intended to stress logic (as in psychology) but the structural discourse of the whole metaphorical process: the image, the feeling, the meaning (i.e., ISM) all together as a single undivided and unified effect. Thus, in rhetoric, sense and feeling have an informative value as does the total metaphor, as does the image, its emotive component, and its meaning in the ordinary sense of the word, since all of these comprise the image or the metaphor in an indivisible fashion. In the tradition of rhetoric, metaphor is clearly a new message. Ricoeur (1979) has tried to differentiate the two different uses of terms by calling one a psychological theory of metaphor and the other a semantic theory of metaphor and concluding that the psychological theory of metaphor is deficient. His argument is based on lack of acquaintance with the psychological theory. Ricoeur (1979) uses such pejorative terms as "mere psychological features" and indeed draws a boundary between the rhetoric theory and psychological theory, an artificial distinction at best.

In the field of rhetoric those theories which deny the cognitive or informative import of the metaphor ascribe a substantive role to the relationship between image and feeling. According to classical rhetoric, discourse aims at persuading and pleasing, a position similar to Kant's theory of aesthetics where pleasure also plays a central role. Although image remains a central concern of these theories, the feeling component is considered an integral part of the process. Ricoeur, who supports an interactive conception of the process, considers that feeling has a semantic bearing and therefore links up with the cognitive on the opposite side of the image. However, at the same time, Ricoeur (1979) tries to follow a literary theory of feeling as if it is something different from the psychological theory. He accepts a more somatic (S) view of emotions, saying that during emotion mental experiences are "closely tied down to the bodily disturbances, as is the case in fear, anger, pleasure, and pain...to the extent that in emotions we are, so to speak, under the spell of our own body, we are delivered to the mental states with little intentionality, as though in emotion we 'lived' in our body in a more intense way" and insists that poetic feelings are different from emotions since they have a specific kinship with language. Ricoeur's theory fails to explain the status of poetic feeling where the vehicle of communication is not language but painting and music. Perhaps Ricoeur's theory can be made more palatable by extending the meaning of language to include nonverbal language, but this blows apart Ricoeur's fundamental distinction between literary and psychological theory on the basis of distinction between feeling and emotion. It would be bad metaphysics to throw emotion out of the field of literature and aesthetics. In music the use of the word "emotion" seems as natural and profound as the use of the word "feeling." Emotions, just like feelings, are linked to images (I) and meaning (M), as is cognition. Ricoeur (1979), in fact, admits that the tendency to identify emotion and feeling exists within literary criticism when he complains, "we keep applying to feeling our usual interpretation of emotion". If we decide to define feeling as a second order intentional structure (Strasser, 1956) within the "S" structure, we can perhaps avoid confusion. Then we can allow feeling the status of complex intentionality without jeopardizing its inherent link with emotion.

The necessity of positing in literary criticism a separate status for feeling stems from the obvious fact that thinking in its so-called objective form is always divorced from emotion and this, to the literary mind, does not represent the true status of reality, or even the essence of thought. Thought in objective sciences must be unemotional, but in the psychological science, the separation of thought from emotion is at best artificial and undesirable, even inhuman. Feeling abolishes the distance between image and thought, between the known and the knower. In the ISM formulation, the feeling and emotion "S" stand between image "I" and the meaning "M." The "S" can only be placed in the middle and nowhere else in the triangular relationship, neither on the outer end of the "I" nor on the other far side of the "M." If we do so, we will bring the "I" too close to "M," and we will have as a result a cold intellectualized linking of "I" and "M" which must be avoided.

The basic minimal unit of psychological experience which involves imagery is the ISM sequence. The interconnected operations can be found in this three-dimensional unity which is comprised of a vivid image (I), a somatic or body response (S), and a meaning (M). It is important to note that normal imagery experience tends to occur mostly in ISM order. The image is accompanied by a somatic or body response (S) which is always a specific type — skeletal, proprioceptive, motor-neural impulse, sensory experience, etc., and a measure of meaning.

Sometimes the ISM does not occur in its proper order for various reasons, some of which are discussed next. There are six basic operational variations of the ISM — ISM, IMS, MIS, MSI, SIM and SMI. Brief examples of these variations follow.

ISM. In this sequence the image (I) is followed by experience of body feelings or somatic response (S), followed by meaning (M) of the experience. Example: A person sees a snake in the path in front of him; he stops walking, looks around, and thinks of an alternative route. The ISM is the most common and the most natural order of experience. Generally speaking, a mental process should function only on the basis of the ISM, because it gives a person a handle on the natural sequence of a space event, his body relationship to it, and the context of meaning in which they coexist. The other operational variations — IMS, MIS, MSI, SIM, and SMI — each stand for certain functional modes which only selectively apply. It is possible for someone to function in any of these variations under special situations. For instance, there are times when people need to think before imagining or feel before imagining or act before thinking, and so forth. However, if an operation in any of these variations becomes fixed or permanent, a malfunction develops.

IMS. In this sequence the image (I) appears first, which is followed by the meaning (M), and then the somatic response (S). Example: A person mistakes a rope for a snake and immediately becomes fearful and runs away. Here the somatic response is initiated through a misunderstanding. The misconception covers up the underlying reality and the appropriate response to the rope. In this variation the image is dominated by the meaning and is stopped from releasing its own meaning and somatic response. Most patients misinterpret reality through this variation, for instance, positive events are interpreted as negative events, loved ones are seen as hostile, etc. Such habitual fixations generate neurotic behaviors. Therapeutically, a patient is asked to see an image, to relax and experience the somatic response before making an opinion. As the somatic response is encouraged to appear before the formation of meaning, the tendency toward formation of misconceptions can be repaired. This leads to restoration of IMS to ISM.

MIS. In this sequence the meaning (M) of the experience is followed by the appearance of the image (I) and then the somatic response (S). Example: A person was expecting a snake in the room, and while entering the room the person sees a rope lying on the floor and perceives it as a snake and runs. In this variation the person's own thoughts may cause him to see reality erroneously and experience a physiological response accordingly. Neurotics perceive reality according to their rigid views. These views suppress the emergence of new perceptions. After the person is opened to new experiences in imagery, the old belief systems are discussed and erased in brief interludes. As a result, new thinking develops.

MSI. In this sequence meaning (M) is followed by a somatic response (S), followed by an image (I). Example: A person believes that there is a snake in the room and approaches the room with tension, and reacts to the rope in the room as if it were a snake. In people, thinking is translated into somatic states which, in turn, generate imagery. People who think that the world is threatening have a fearful life and see situations as always hopeless where others would see hope. This type of person should be taught to respond to selected positive images through relaxation and positive body experiences.

SIM. In this sequence somatic response (S) is followed by image (I) and then meaning (M). Example: A person who is drunk is told that there is a snake in the room. He sees one accordingly and runs out. The person in physical tension or during muscle relaxation sees images accordingly. A physically altered condition causes a person to see images differently and according to the altered state; thus, drugs can influence perception. Over a long period of time neurosis changes the body equilibrium and the result can be neurotically created physiological illness. Relief comes through muscle relaxation and introduction of images which open up new avenues for body expression. During deep body relaxation spontaneous emergence of imagery experiences has often been reported. This can be actively encouraged through playing of music and introduction of other similar aids.

SMI. In this sequence somatic response (S) is followed by meaning (M) and then image (I). Example: A person who is acutely physically tired or is under duress is given an idea which he finds difficult to resist and, as a result, he sees images accordingly. A person in a submissive emotional state is vulnerable to acceptance of a suggested reasoning and its corresponding images. In hysteria a forgotten episode may cause pressure on consciousness through its attached somatic aspect. The patient may experience the somatic aspects, even vaguely feel its meaning, but since the visual image of the event is not accessible to consciousness the event remains repressed. The treatment requires recovery of the original image and experience of the somatic response and meaning in the regular ISM order. The restoration of the totality of experience in its natural order leads to normalization of consciousness.

In the ISM variations described above we find another covert phenomenon not noticeable at the surface. M, which is the meaning in

ISM, is an experiential mode which may or may not express itself in a verbal form, but has a definitive although as yet unrealized access to overt language. But a word always has more associations than may be intended in a specific context. As a result, when a word represents a meaning we have two simultaneous processes which converge as well as diverge and this condition represents two separate language and imagery links with their two separate imagery and linguistic chains. Thus, a covert and an overt imagery process goes on at each verbal point, as a covert and an overt verbal process goes on at each imagery point. One reaches the conclusion that words, which are usually found at the fringes of the imagery process, have something to do with the imagery process, just as the images which are spotted at the fringe of verbal expression have something to do with the verbal process. The experience tells us that these fringe events, whether imagistic or verbal, are capable of creating a whole new line of imagery experience or verbal thought which clearly diverges from a convergent focal mental process. The meaning of meaning is thus made more complicated by the existence of fringe events. This complication, when aided by mechanics or by an enactive process, sparks off many new possibilities. Here we have a very dramatic but perfectly logical relationship between conscious and unconscious, between experience and term, and between term and word in the general flow of experience.

The mind at various levels, whether you refer to the "unconscious" as the "conscious," or "consciousness" separated from "unconsciousness," or "consciousness" which thwarts the "nonconscious," represents a complex set of implications. In this light, the mind is represented by a family of terms. Considering the nature of structure represented by each separate term, all separate spheres of terms reflect a kinship of relationships both convergent and divergent. Since any term suggests other terms variously related at convergent as well as divergent levels, the glimpsing of mental structures is like envisioning of possibilities. In this way the implicit and the explicit merge as well as goad each other on, which is the process of making connections. But a connection in this sense is not an arbitrary or verbal association as in free association but an authentic, firm and bound relationship which connects two or more valid terms by uncovering the underneath experiential process. Two terms are required in logic, and a minimum of three terms in adventure as has been revealed in the evolvement of drama in Greece when finally a third actor was introduced to enrich the two actor setting. The meaning is not composed of two terms only but three, two original images and the third big spark which is the metaphor itself. One might say that a psychological experiment is also a metaphor aimed at revealing the original face of reality.

2. Vividness Unvividness Paradox

As mentioned earlier Ricoeur distinguishes between the psychological theory of metaphor and the literary theory of metaphor. His argument sows the seeds of dissension between psychology and literature, since this division at least from the psychological side is unsupported. To elucidate this, one would want to pose to the literary critic the question: Are literary goals unpsychological? In fact, the strategies which are sought in literary metaphor are empirical and would not be possible if, in some way, the literary metaphor did not observe the laws of the mind. The literary theory of metaphor as formulated by Ricoeur has another serious problem with it, a problem which has been the center of controversy in the psychological theory of imagery, namely, the mechanics of vividness in an image. According to Ricoeur's theory of metaphor, two images are struck together to create a metaphor, which is another illumination, another vividness. But there is another quality which metaphors display, and that important quality is not clearly represented in Ricoeur's theory — subliminal imagery. Metaphors are known to offer clear as well as subliminal messages, a series of vague images attached to the main clear image. Ricoeur's omission of this dimension concerning metaphor provides us with a departure point for discussion of a central controversy around vividness of imagery in psychological research. David Marks' (1972) work in psychology, which concerns the role of vividness, is the subject of much controversy in this direction.

The clear status of vividness in imagery studies is hedged in by many issues and considerations. In psychological studies the relationship between imaginal mediators and cognitive performance has been pretty much established as correlational, but the search for a direct causal mechanism linking vividness of imagery and recall, however, continues on. There are many complex features to this loaded issue, the most central being the very meaning of the word "performance" which appears so often in laboratory evidence. By performance is generally meant memory recall or achievement of a certain skill, outputs in which functionality of imagery vividness is measured and reported. This meaning of performance then is generalized into imagery's relevance or irrelevance to mental functioning. In the development of this view the relationship between phenomenology and function is further put under the stress of a confusing and overly generalized assumption that vividness of imagery is the sole test of functionality of imagery and if vividness has flunked this test the case for imagery is lost. However, it is well known that imagery occurs profusely when a muscular task such as riding a bicycle is being learned, but after the task is learned the stored skill excludes mental imagery. We can expect a role for imagery in formulation of muscular skills, arithmetic, verbal habits and so forth. But the absence of imagery in the recall of these skills does not prove that imagery was not originally used in the formation of these skills. Furthermore, learning to perform some special muscular skill such as throwing of darts at a target may really involve practice of perceptual imagery in the presence of an actual target than imaginary throwing of darts at an imaginary target. There could be special attributes of various skills which need specialized connections with imagery. Most experiments on imagery have shown a consistent tendency to avoid the complexity associated with vividness and involve only a copy role of vividness. Thus, there is always a tendency toward confusing memory with imagination, or in an opposite way excluding the play of imagination in memory tasks by overdefining memory tasks. In a way plasticity is an important aspect of vividness in imagination imagery, but plasticity vanishes in fixed images generally experienced by neurotics. Thus, vividness would reflect plasticity or lack of plasticity in the light of how we originally formed the vivid image or toward what purpose we are using it. The process of recall offers some features which are common with imagination, and testing may not show correlation if the given task avoids imaginative features. Unavailability of an image is, in fact, a symbol in the imagery process. This suggests that vividness needs to be redefined in a paradoxical way such as: An image which is not vivid or is absent serves a different function than that of a vivid or a present image, so that vividness or unvividness are two functional attributes of the same vividness phenomenon. In brief, unvividness is functionally vividness. The psychodynamic importance of this paradoxical definition of imagery can be emphasized through what has often been observed in the clinic. A certain type of patient suffering from defensiveness does not see images vividly during the session, but begins to see vivid imagery the moment he or she leaves the clinic. Another type of patient generally shows excellent spontaneous imagery response, but the moment the therapist introduces an image the patient responds by complaining that the image is very vague. These types tend to withhold the very thing they are expected to give. In brief, they are self-denying or denying to others or, in other words, punishing in their life style. How else can one explain this selective unvividness on their part since it comes right at that point when a demand for imagery is made? This belligerent type perhaps exists even more widely outside the clinic. In the student community it can be considered even rampant in the sense that students do frustrate psychological experiments in many ways. Without being conscious of it, they may go along with the expectation of the experimenter or go against it, a kind of paradoxical demand effect. I am not claiming that the students cheat or do this maliciously, but that the desire to comply or to deny is a normal propensity and that the desire to deny may be more prevalent in the student population since this population happens to be very testy and competitive and likes to play games. Students belong to that adolescent age level when contrariness is a normal feature of development. In my own judgment, experiments on imagery must be constructed first involving this paradoxical phenomenon relating to what might be called imagery games around vividness.

Thus, we find that unvividness is functional vividness in the sense that it describes a mode of operation which cannot be performed in any other way. If we put vividness where only unvividness works, we would not get the results we expect. So far we were talking about the role of unvividness concerning a single image which could be seen or missed in one glance. Here is a sequential example in which the report of unvividness of an image is followed by vividness of the same image, and the two are separated by an intervening somatic event. A mother who lost her baby many years ago reports a very vague image of the baby on first trial, but after concentration, she starts crying and then reports the image of the baby to be very, very vivid. Crying was in the middle of unvividness and vividness, and moved unvividness to vividness. It is, however, not only a whole single contemporaneous, or sequential, set of two images that may show paradoxical behavior involving vividness, but the same whole single image may show characteristics of vividness and unvividness pertaining to various parts of the manifestation. The mother still cannot see the legs of the baby whose image otherwise is now vivid. However, the concentration on the hands of the baby leads to the vividness of the legs. Furthermore, not only does a vivid image make another unvivid image vivid, but also, contrarily, an unvivid image can make another vivid image even more vivid. Thus, unvividness in one part of the image may contribute to enhancing vividness in the other part and in that sense the first unvividness is a form of vividness, but we have got to call it unvividness because otherwise it will be confused with vivid-vividness. You have got to have a word for a phenomenon, and once we do, we are stuck with it.

We can see why a paradoxical statement concerning vividness becomes not only plausible but essential. In nonverbal arts, the vagueness of an image in a covert position is its vividness. It is well known that, in visual art, images which are peripheral to the focus are always rendered vague. The unclear image in the periphery of a painting is a good image and that is the vividness it should have and no more. The lack of vividness is thus the functional quality of the peripheral image rather than its absolute value and this functional quality relies entirely on the role it is expected to play in a context. The psychological importance of this can be presented through a brief experiment, in which subjects were asked to image their parents standing in front of them and report the left-right position of the parental images (Ahsen, 1972, 1977b, 1981). Most subjects are able to see the parents' images clearly in some kind of fixed left-right position, but some do not. One of the subjects, R. D., for instance, reported: "My mother is standing on the left and her image is very clear. My father is on the right and his image is very vague. It is a recent image from the time when my father developed cancer and died a year later." However, when R. D. was asked to recall his father before he developed cancer and then see his parents' images in left-right positions, he reported: "Now my father is on the left and the image is very clear. He appears very happy and playful in the image. My mother appears on the right. I date the image about 15 years back." The sudden shift in the vividness of the father's image from very vague to very clear was accompanied by a switch in the left-right position, and relegation of the image to a distant time frame. This reported change in vividness has broken all the rules which would be expected in a normal study of correlations between vividness and recall. Recent images should be brighter than images from the remote past, etc. However, the vague image of the father is vague *because the subject does not want to see it.* R. D. said, "Near his death he looked so emaciated and so horrible that I do not want to see him again in that state." Marks (see Chapter 10) and Molteno (1984) report interesting experimental studies of this phenomenon (Ahsen, 1972, 1984d). However, the main point being made here is that vividness and recall are causally connected. In fact, the vividness-unvividness dimension is an inseparable part of the image and in this paradoxical tie the image does not embody pure recall, rather embodiment of the attitude toward recall. In brief, the image is not a memory, but a symbol. Any attempt which employs memory methods for the study of imagery in a major way, such as Paivio's studies (1972), would be confusing the imagery process as a whole with memory. Memory is the worst criterion one can expect to inject into the study of imagery and leads serious problems in understanding the function of imagery mechanics in the flow of mental processes and their relationship to vividness.

The argument concerning vividness can be further elucidated through a similiar problem which theoreticians have come across concerning hypnosis. Ernest Hilgard (1977) is critical of the state theory of hypnosis as a trance, and relating hypnotizability even to personality. One thinks in a parallel way of a state theory of vividness or relating vividness to personality, very much a similar issue. Vividness is almost treated like a trance in general imagery theory. This common tie between vividness of imagery and hypnosis is clearly illustrated by Josephine Hilgard (1970) who discovered that the capacity for imaginative involvement was important for hypnotic susceptibility. In this respect she emphasized the important role played by enjoyment of the feelings of the moment, escape from the world of reality, a time-limited experience of adventure, active involvement and childlike enjoyment of excitement, power and sense of triumph. Studies that have variously linked vividness, recall and manipulation show that vividness does not itself cause recall; in fact, various degrees of vividness accompany recall during various manipulations, and it depends how relevant, interesting or adventuresome the manipulation is. In brief, we cannot evoke vividness as a trancelike value in itself. The degree of vividness or unvividness has to be embodied in a particular image as a consequence of the type of manipulation which accompanies it. The fixed idea that vividness of imagery is the likely basis for all types of recall as well as all forms of imagination is considered by Ernest Hilgard (1977) to be symptomatic of the old reductive psychology. For instance, Gordon (1950) has emphasized the ability to control images by taking into account something beyond sheer vividness. In fact, creative sculpting of an image involves steps starting from something which is vague but nevertheless there, and making that apparently vague image vivid. Hilgard insightfully remarks that Marks' VVIQ in this very aspect differs from Sheehan's (1967) briefer version of Betts' Questionnaire Upon Mental Imagery (1909). Marks' VVIQ provides "some manipulation of the images involved, which moves it into the territory of creative imagination. For example, a storefront is pictured from across the street, and the items in the store window are viewed as though the viewer had moved in closer. Next some particular item is examined as though it were in full attention. By scoring vividness at each of these points of manipulation something is added to mere vividness of visual imagery as such" (Hilgard, 1977, pp. 103-104).

Thus, the type of operational dynamic involved in imagery experience and not merely memory recall is the key to the vividness paradox. Marks' VVIQ is an excellent test which provides a fair mix of the mechanical and the dynamic to show the efficacy of the vividness in performing mental tasks which resemble life. Marks' VVIQ takes us beyond the self-report issue into the dynamics and mechanisms involved in these self-reports and proves that vividness and performance do correlate when these dynamics are accounted for.

Thus, the real argument in imagery is not finally for epistemological primacy of self-reports, but for the importance of what can be called *introspective operationalism,* based on introspective monitors, and the knowledge of these life monitors through empirical evidence. The imagery reports are not used for mere evidence of what the verbal communication is saying but description of actual effects that can be procured in a living and practical way for guidance and control of mental phenomena. In the words of Alan Richardson (1984), "It is a contorted form of logic that forces some psychologist to say, 'All I know is the words I have heard,' as if the referents of these words were psychologically unimportant" (p. 122). In a similar light, J. T. E. Richardson's (1985) defense of imagery based on convergence of operations sounds reasonable, but this position too is marred by its unnecessary associations with a linguistic view of imagery drawn from Wittgenstein. The convergent approach to imagery is useful but, at the same time, we also need a divergent approach which can involve the other more generative side of imagery operations which comprise more of what the essence of working with imagery is all about. We do not need a linguistic defense of imagery when we have an imagery based defense of imagery. Once and for all, we need to go beyond Pylyshyn's propositional view of imagery and base imagery research firmly in the introspective area, both in the experimental laboratory and the field study of imagery effects. We are looking forward to the advent of what I call New Structuralism in psychology.

V. CONCLUSION

Locke failed to fully elaborate and exploit the possibilities of an empirical psychological science that could have involved straightforward introspective material as much as the study of imagery mechanisms, myth and legend. In our discussion we found that the commonsense view of things, rudimentary reflection, ritual reflection, myth reflection, scientific reflection and political reflection were truly interdependent. We drew from this intensive view of interactive reflections an empirical theory of mind which glimpsed a very expansive view of a truly complex phenomenon, namely, imagery. The presentation embraced a convergent and a divergent view of imagery operations, providing us with a new structural vision of an enriching reflection which was not alienated nor artefactually bred within the consciousness system. Modern imagery theory and practice provides such a construction of consciousness.

REFERENCES

Ahsen, A. (1965). *Eidetic psychotherapy: A short introduction.* New York: Brandon House.

Ahsen, A. (1968). *Basic concepts in eidetic psychotherapy.* New York: Brandon House.

Ahsen, A. (1972). *Eidetic Parents Test and analysis.* New York: Brandon House.

Ahsen, A. (1977a). *Psycheye: Self-analytic consciousness.* New York: Brandon House.

Ahsen, A. (1977b). Eidetics: An overview. *Journal of Mental Imagery, 1,* 5-38.

Ahsen, A. (1981). Imagery in hemispheric asymmetries: Research and application. *Journal of Mental Imagery, 5,* 157-194.

Ahsen, A. (1983). Odysseus and Oedipus Rex: Image psychology and the literary technique of consciousness. *Journal of Mental Imagery, 7*(1), 143-168.

Ahsen, A. (1984a). *Trojan horse: Imagery in psychology, art, literature and politics.* New York: Brandon House.

Ahsen, A. (1984b). *Rhea complex: A detour around Oedipus complex.* New York: Brandon House.

Ahsen, A. (1984c). Reading of image in psychology and literary text. *Journal of Mental Imagery, 8*(3), 1-32.

Ahsen, A. (1984d). Heartbeat as stimulus for vivid imagery: A report on individual differences and imagery function. *Journal of Mental Imagery, 8*(2), 105-110.

Ahsen, A. (1984e). ISM: The triple code model for imagery and psychophysiology. *Journal of Mental Imagery, 8*(4) 15-42.

Allport, G.W. (1924). Eidetic imagery. *British Journal of Psychology, 15,* 99-120.

Bandura, A. (1977). Self-efficacy: Toward a unifying theory of behavioral change. *Psychological Review, 84,* 191-215.

Berkowitz, L., & Donnerstein, E. (1982). External validity is more than skin deep. *American Psychologist, 37,* 245-257.

Betts, G.H. (1909). *The distribution and functions of mental imagery.* New York: Teachers College, Columbia University.

Bloom, H. (1982). *The breaking of the vessels.* IL: University of Chicago Press.

Breton, A. (1962) *Manifestes du surrealisme.* Jean-Jacques Pauvert, n.d. (Original work published 1924)

Buckler, W.E., & Sklare, A.B. (1966). *Essentials of rhetoric.* New York: Macmillan.

Bugelski, B.R. (1977). Imagery and verbal behavior. *Journal of Mental Imagery, 1*(1), 39-52.

Bugelski, B.R. (1982). Learning and imagery. *Journal of Mental Imagery, 6*(2), 1-92.

Burke, K. (1957). *The philosophy of literary form.* New York: Vintage Books.

Cautela, J.R. (1977). Covert conditioning: Assumptions and procedures. *Journal of Mental Imagery, 1,* 53-65.

Cohen, J. (1966). *Structure du language poetique.* Paris.

Count-van Manen, G. (1985, October). *George Herbert Mead on mental imagery: A neglected nexus for interdisciplinary collaboration with implications for social control.* Paper presented at the 9th American Imagery Conference, Los Angeles, CA.

Dennett, D.C. (1981). Two approaches to mental images. In N. Block (Ed.) *Imagery.* Cambridge: MIT Press.

Descartes, R. (1966). *The meditations and selections from the principles of Rene Descartes.* (John Veitch, Trans.). La Salle, IL: Open Court.

Descartes, R. (1966). *Descartes: Philosophical writings.* London: Thomas Nelson & Sons, Ltd.

Frege, F.L.G. (1978). Sense and reference. In P. Ricouer (Ed.) *The rule of metaphor: Multidisciplinary studies of the creation of meaning in language.* Ontario: Univ. of Toronto Press.

Goodman, N. (1968). *Languages of art: An approach to a theory of symbols.* Indianapolis, IN: Bobbs-Merrill.

Gordon, R. (1962). *Stereotypy of imagery and belief as an ego defence.* London: Cambridge University Press.

Hebb, D.O. (1968). Concerning imagery. *Psychological Review, 75*(6), 466-477.

Heidegger, M. (1967). *What is a thing?* Indiana: Gateway Editions.

Hester, M.B. (1967). *The meaning of poetic metaphor.* The Hague: Mouton and Co., N.V., Publishers.

Hilgard, E.R. (1975). Neo-dissociation theory: Multiple cognitive controls in hypnosis. In L. Unestahl (Ed.), *Hypnosis in the seventies.* Orebro, Sweden: Veje Forlag.

Hilgard, E.R. (1977). *Divided consciousness: Multiple controls in human thought and action.* New York: Wiley Interscience.

Hilgard, J.R. (1970). *Personality and hypnosis: A study of imaginative involvement.* IL: University of Chicago Press.

Husserl, E. (1958). *Ideas.* W.R. Boyce Gibson (Trans.) New York: Macmillan.

Jacobson, E. (1962). *Selected writings.* 2 vols. The Hague: Mouton and Co., N.V., Publishers.

Janet, P. (1889). *L'Automatisme Psychologique.* Paris: Felix Alcan.

Kakar, S. (1982). *Shamans, mystics and doctors.* New York: Alfred A. Knopf.

Kaufmann, G. (1980). *Imagery, language and cognition.* Bergen: Universitetsforlaget.

Laing, R.D. (1965). *The divided self: An existential study of sanity and madness.* Baltimore: Penguin Books.

La Mettrie, J. de. (1961). *Man a machine.* La Salle, IL: Open Court.

Levi-Strauss, C. (1964). *The raw and the cooked: Introduction to a science of mythology: 1.* New York: Harper Torchbooks.

McDougall, W. (1926). *An outline of abnormal psychology.* London: Macmillan.

McKellar, P. (1979). *Mindsplit: The psychology of multiple personality and the dissociated self.* London: Dent.

Marks, D.F. (1972). Individual differences in the vividness of visual imagery and their effect on function. In P.W. Sheehan (Ed.), *The function and nature of imagery.* New York: Academic Press.

Matthews, J.H. (1982). *Surrealism, insanity, and poetry.* New York: Syracuse University Press.

Mead, G.H. (1934). *Mind, self and society from the standpoint of a social behaviorist.* IL: The University of Chicago Press.

Mead, G.H. (1938). *The philosophy of the act.* IL: The University of Chicago Press.

Molteno, T.E.S. (1984). Imagery in the Eidetic Parents Test. M.Sc. thesis. University of Otago, New Zealand.

Orne, M.T. (1962). On the social psychology of the psychological experiment: With particular reference to demand characteristics and their implications. *American Psychologist, 17,* 776-783.

Paivio, A. (1972). A theoretical analysis of the role of imagery in learning and memory. In P. Sheehan (Ed.), *The function and nature of imagery.* New York: Academic Press.

Paivio, A. (1976). Images, propositions, and knowledge. In J.M. Nicolas (Ed.), *Images, perception, and knowledge.* (The Western Ontario Series in the Philosophy of Science.) Dordrecht, The Netherlands: Reidel.

Pylyshyn, Z.W. (1973). What the mind's eye tells the mind's brain: A critique of mental imagery. *Psychological Bulletin, 80,* 1-24.

Richardson, A. (1984). Strengthening the theoretical links between imaged stimuli and physiological responses. *Journal of Mental Imagery,* 8(4) 113-126.

Richardson, J.T.E. (1985). Subjects' reports and converging operations. *Journal of Mental Imagery,* 9(2).

Ricoeur, P. (1979). The metaphysical process as cognition, imagination, and feeling. In S. Sacks (Ed.), *On metaphor.* Chicago: University of Chicago Press.

Sartre, J.P. (1978). *The psychology of imagination.* London: Methuen.

Sheehan, P.W. (1967). A shortened form of Betts' Questionnaire Upon Mental Imagery. *Journal of   Clinical Psychology,* 23, 386-389.

Spicker, F.S. (1970). *The philosophy of the body.* New York: Quadrangle New York Times Books.

Strasser, S. (1956). *Das gemut.* Freiberg.

CHAPTER **2**

Imagery and the Unconscious

Peter McKellar

Who save the madman dares to cry "Tis I am right, you all are wrong"?
"You all are right, you all are wrong," we hear the careless Sufi say,
"For each believes his glimmering lamp, to be the gorgeous light of day.
Thy faith, why false, my faith, why true? Tis all the work of Thine and Mine,
The fond and foolish love of self, that makes the Mine excel the Thine."

THE KASIDAH

Haji Abdu Al-Yazid
(Sir Richard Burton)

I. INTRODUCTION

The poem is a plea for thoughtful consideration of alternative ways of thinking. It carries a message for theorists within psychology. Rival systems of thought illuminate different things, but the preferred constructs they embrace may obscure others. The poem stresses a resemblance between monomania in deluded pathological thinking, and dogmatic one-sidedness-quasi-pathological stability - that sometimes finds expression in the sane. Richard Burton wrote the poem under one of his Eastern noms de plume. He had just returned form his visit, disguised as a Moslem pilgrim, to Mecca. He uses a vivid metaphor. In this chapter I shall be much concerned with metaphor, with analogies that play a part in psychological theory and with orientations that result from thoughtful examination of alternative formulations.

"I wish he would explain his explanation," complained a critic of a writer whose poetry was considerably more obscure than Richard Burton's. To "explain" in this sense is to re-state what is being said by putting it into different words. Perhaps a different analogy will be used. Ability to reformulate in the interests of comprehension is important in psychological theory, no less than in understanding poetry. And once he has forced himself to reformulate, a given theorist may himself achieve better understanding of what he is, and is not, trying to communicate. Moreover, psychologists seem to have a well-established habit of often saying very much the same thing in different words. Certain words have become the banners of schools of thought, often defended with some degree of militancy. Today, as in the past, rival theorists have lacked enthusiasm for "translating" favored labels, theories and models. A sign of hope. It is a vigorous growth of interest in the uses of analogy - simile and metaphor - in human thinking. Thus, Julian Jaynes in the 1970's drew attention to "the metaphor language of mind." He showed how people repeatedly use analogies drawn, for example, from the structure and functioning of the body, when trying to introspect about consciousness (Jaynes 1976). Ortony (1979) has edited a book devoted to the relations between metaphor and thought. To this book Paivio makes a contribution. As he puts it, "metaphor is a solar eclipse": it both enlightens and obscures. Metaphors "block out the central stuff so that you can see the subtle stuff better" (Paivio, 1979, p.169). Metaphors merit attention in theories about imagery the unconscious, and the dissociated.

II. METAPHOR, ANALOGY, AND THEORY

The *Oxford English Dictionary* defines metaphor as "application of a name or descriptive term to an object to which it is not literally applicable." Five decades ago the psychologist Spearman (1930) examined the relationship of figurative language to creative thinking. He argued that we observe a system of relations in one context, abstract it, and apply it to a different subject matter. Analogies can assist exposition, and sometimes explanation. From the linguistic side Buckler and Sklare (1966) make a similar point: "We may use an analogy to make something unfamiliar clearer by comparing it with something familiar" (p.182). Many examples of this can be found in both psychoanalysis and contemporary imagery theory. Breuer and Freud made considerable use of analogies. They likened the unconscious to many different things: and no less than 23 such analogies are listed, as such, in their index. These include a buried city, a building of several stories, and a surgical operation (Breuer & Freud, 1893-1895/1955).

They also write of a foreign body in living tissue, movements of a knight in a chess game, unlocking a door, decoding pictographic script, and of a defile of consciousness.

In making substantial use of analogy the pioneers of psychoanalysis do not seem to have differed greatly from later theorists. Consider, for example, imagery. A book edited by Block (1981) reports "a debate" between those who view images as "pictures in the head," and others who liken them to the "symbol structures" of modern computing technology. "Icons" figure prominently in this debate (in *both* Bruner's and Neisser's different senses), as do their internal auditory equivalents, namely, "echoes." Both terms are metaphors. After making a long excursion into some imaginative anthropology, Dennett (1981) contributes by introducing the labels "iconophiles" and "iconophobes" as names for the rival theorists. With the present surplus of analogies - seemingly rather disembodied entities, as it were, "floating about" in psychology - a contribution elsewhere by Pinker and Kosslyn (1983) seems timely. They remind us that mind is a biological phenomenon. Their account provides systematic exposition of imagery theories and also a classification. And they give guidance in choosing between alternatives in terms of both neurological and behavioral criteria. Also timely is the contribution by George Miller (Ortony 1979), who knows a great deal about how psychologists use words. He suggests that the earlier concept, "apperception," is a convenient superordinate term to embrace much of the activity and thought about information processing being reported in the current journals. Like Miller, I shall be interested in relating current theories, analogies and concepts to those of an earlier period. These include "the unconscious" and "personality dissociation," both with their own rich accumulation of analogies and metaphors. Re-examination of these two concepts may help to broaden the often too narrow front of the present arguments about imagery. Despite Pylyshyn's (Block, 1981) title "The imagery debate," there is not just one such debate. There are, and should be, many.

III. IMAGERY AND THE UNCONSCIOUS

"Imagery" as such has been remarkably absent from psychoanalytic theory. Nevertheless, it is very much there, referred to in other words. The noms de plume used are not confined merely to "fantasy," and the name of the journal, *Imago*. Consider three very basic psychoanalytic concepts: repression, free association, and transference. One of the things we image is ourself. In discussing *repression* - and return of the repressed - Freud was very much concerned with the past doings of the imaged self, and strong motivation to forget them. Again, Freud's "concealing memories" embrace what others may call "images" - whether actual memories or imaginings - often involving the doings of the self. Likewise, the process of *free association* relates to imagery of past, present and future. Moreover, some of the forms of resistance to such association involve imagery. An example is a dream that the patient brings to an analytic session; he is ready to talk about it, and uses it as a way of avoiding having to associate about other things. A wise analyst may respond by saying "What is it about your feelings toward me today that you don't want to talk about?" Consider *transference*. This may involve the patient's belief, or half-belief, that the analyst sitting behind him, out of sight, has changed into somebody else. The analyst is very often imaged as a parent figure. Finally, even imagery psychologists sometimes forget that probably the most influential book that has been written on imagery had as its author Sigmund Freud. *The Interpretation of Dreams* is about one major type of imagery, and it also deals more incidentally with another - the imagery of the hypnagogic state.

Among non-Freudian systems of depth psychology, again we find much concern with imagery. Techniques of dream interpretation and of active imagination have been important features of Jung's analytical psychology. Jung drew his analogies and models from Eastern religions, alchemy and astrology, and made much of the notion of projection. Thus, in the case of alchemy, psychological processes - often imagery - can intrude on the chemical realities. Again, in Jungian versions of transference, the imagery which complicates the relationship is not confined to memories from the life history of the individual. Also, for Jung it involves archetypal imagery from the "collective unconscious." The relation of this both to Freud's "Id" and to Groddeck's (1928) concept of the "It" offers a fruitful area for terminology translation. Foremost among the academic psychologists who sought common ground between Freud and Jung was Henry Murray. Murray's Thematic Apperception Test provides another technique for personality study, very largely through visual and auditory imagery. Take another instance, involving Alfred Adler, and his technique of personality study through seeking the early memories of that patient. Similarities between the imagery involved and Freud's concealing memories again offer a field of inquiry for terminology translation.

Since the same word may refer either to perceiving or to imaging, I will introduce the convention of using italics for the imagery activity. Thus in dreaming or hallucinating a person confuses his imagery *seeing* or *hearing* with real perception. Conversely, in the Perky phenomenon what is actually a visual percept is misinterpreted as *seeing* in the imagery sense. The many things people can *do* in their imagery is not confined to quasi-sensory activities. In waking imagining, dreamlife and the intervening hypnagogic state, interesting violations of the laws of physics occur (McKellar & Simpson, 1954). As regards dream imagery, I pointed out some years ago (1957) that "in dreams all things are possible," and moreover we as dreamers "accept without surprise or alarm deviation from these laws." Yet, when awake we rely on them with confidence. An example is overcoming gravity and "flying like a bird." The pioneer investigator, Mary Arnold Foster, became skilled in flying in her dreams. As regards hypnagogic phenomena, considering the case of a Professor Newbold who, while falling asleep found himself "flying face downwards about 20 feet above the ground." He reported "It is always night and I am following a road, trees, fences, fields dimly seen along the roadside" (Leaning, 1925, p.151). It is not difficult, I have suggested elsewhere (1957), to envisage the interpretation such *flying* would have received in an earlier witchcraft-ridden age. Obviously, this was transporting oneself to a meeting of witches, as witches *did* in those days. (McKellar, 1957). There remain in the world today many societies known to anthropology in which such supernatural interpretation would seem obvious. In other words, there are still societies and individuals who would not make the distinction I am making between imaged *doings* and real actions.

A characteristic of some, but not all, imagery, is autonomy: it seems to come from "outside" and the imager himself lacks control over its onset and content (Gordon, 1962). Horowitz (1983) prefers the term "unbidden imagery." The concept of "autonomous," as opposed to "controlled," imagery stems from Jaensch (1930). Elsewhere we have suggested a distinction between (a) primary autonomy relating to onset of the imagery, and (b) secondary autonomy, relating to the inability to change or control such imagery once it is in progress (Marks & McKellar, 1982a). To use a different terminology, "autonomous imagery" - waking, sleeping or hypnagogic - overlaps considerably with what I have elsewhere (1963) labeled "A-thinking." Such autistic or A-thinking (from Bleuler's concept of autism) differs from realistic or R-thinking. The interaction of A-thinking and R-thinking may result in socially useful thought products: poems, paintings, musical compositions, even scientific theories. The A-thinking can provide a kind of involuntary "authorship" and the R-thinking a kind of critically evaluative "editorship." From where does such autonomous or unbidden imagery come? The pioneer of imagery research, Sir Francis Galton (1883), wrote of "a presence chamber" of the mind "where full consciousness holds court" and an adjacent "antechamber" (p. 203). Whether we use Galton's terminology or Freud's or prefer to speak of "dissociated systems," the eruption of imagery from "elsewhere" invites a detailed look at the concept of unconscious mental life. The contributions of some of Freud's predecessors may be examined.

IV. THE UNCONSCIOUS BEFORE FREUD

Many have contributed to the notion of unconscious mental life. Important among them have been Leibniz, Kant, Herbart, Schopenhauer, Von Hartmann, and Nietzsche. Ellenberger's *The Discovery of the Unconscious* (1970) makes reference to two alternative "models" of mental life. On the one hand is *dipsychism*, and the notion of duality, (conscious - unconscious), on the other *polypsychism*, and the notion of a cluster of several subpersonalities. I shall be concerned with the polypsychic model later. As regards the dipsychic one-of conscious and unconscious-Ellenberger (1970) assesses the entire nineteenth century as preoccupied with "the coexistence of these two minds and their relationship to each other" (p. 145). In tracing this notion of an unconscious, Ellenberger has provided modern psychology with one of its most important books of recent years. I can only touch more lightly on some aspects of this history. Von Hartmann's major work, *Philosophy of the Unconscious* (1931), begins with a discussion of the contribution of his predecessor, Immanuel Kant. Hartmann refers to Kant's discussion of "ideas we have without being aware of them" and his conclusion that this notion involves no contradiction in terms. As Kant puts it, "ideas is the genus - under it falls the idea accompanied by consciousness" (Hartmann, p.21). Also it is noted that the original German of Kant implied the notion of "representation." Some difficulties of translation into English occur and there is one I want to take up concerning the work of Schopenhauer who was, like Hartmann himself, much influenced by Kant. For a long time (through the translations of Haldane and Kemp), Schopenhauer's major work *Die Welt als Wille und Vorstellung* was known, in English as *The World as Will and Idea*. A later translator, E.F.J. Payne has provided instead the title *The World as Will and Representation* (1958) As this translator points out, the previous use of the word "idea" for "Vorstellung" failed to bring out the meaning of "Vorstellung in the sense used by Schopenhauer" (p. ix). It may be noted, in passing, that Ellenberger agrees, in that he uses Payne's title. Payne himself writes "in the present translation 'representation' has been selected as the best English word to convey the German meaning." He adds that the word when used by Schopenhauer refers to complicated physiological processes in the brain "the result of which is the consciousness of a *picture* there" (p. ix, italics his). To this excursion into translation I add a comment of my own. Many have recognized that Schopenhauer's "will" largely anticipates the Freudian, and perhaps even more, the Jungian notion of "the unconscious." In other words, the theme of Schopenhauer's major work might best be interpreted as dealing with "The unconscious and its representation." He was much concerned with the images and constructs we build up about unconscious mental life.

A somewhat different orientation to psychoanalysis and later systems of depth psychology may result from this interpretation. Modern science is strongly Kantian in the way it has developed. It is concerned with the constructs we evolve as a means of understanding the phenomena of nature. Physics and biology offer many examples. Freud's predecessor, Schopenhauer, was considerably more Kantian than earlier translations into English implied. As his successors, Freud and Jung followed this tradition. The unconscious is a construct: there is no such place. Nor are ego, id and superego, "little men" quarreling with one another within the central and autonomic nervous systems. They are constructs - abstract ways of referring to complex phenomena. And they carry with them constant temptations to reify. The same temptations are present in the realm of imagery. Although we constantly use this analogy, the mind does not possess an "eye," as Marks (1983) reminds us. Analogies - whether with plumbing, as in the "flow" of libido, or with the programming software of modern computing - may be legitimate. They can also be misleading if we lose sight of the difference between what is being said figuratively, and what is meant literally. Consider geography. Geographers, like psychologists, use constructs. Examples include lines of latitude and longitude, the Equator, and the Arctic Circle. Modern jet travel provides problems with time zones and the International Date Line. One can arrive at one's destination in a sense, "before" departure. And the question "What is the time?" during an International flight may have more than one answer. In other words, geographical constructs, while useful, have their limitations. Within psychology the issue which concerns us is the use and limitations of constructs - Freudian, pre-Freudian, and post-Freudian - relating to "the unconscious."

Freud was a vigorous model builder: indeed, an influential pioneer of this area of human thinking. Men in ancient times, and later, looked at the stars and found ways of grouping them, drawing from models they observed on the earth: the ram, the bull, and so on. Interestingly enough, many of the constellations reflect Greek mythology. In seeking to understand what interested them, Freud and his successors used similar sources. Consider Spearman's analysis, cited above, of abstracting a system of relations and applying it to a different subject matter. In psychoanalysis we likewise encounter many Greek myths and legends:

Narcissus; the Sophoclean version of Oedipus; Mycenae and Electra; and the doings of Prometheus. Flugel (1945) later contributed an interesting insight into one aspect of human motivation from the story of Polycrates, and the "Polycrates complex." The theme: "all is going too well, what disaster is going to follow?" The depth psychologists have used many analogies - often classical in origin - to express their insights into child development, aggression and self-destructiveness, and motivation in its subtler aspects. I refer again to Paivio's metaphor about metaphor itself. It helps us "to see the subtle stuff better." It may be noted that the psychoanalysts had their predecessors in this activity. For example, Nietzsche's Appollonian versus Dionysian typology in *The Birth of Tragedy* had both classical origins and subsequent applications in anthropology.

Some of the lesser critics of psychoanalysis appear to have read few, if any, of its texts: if they have, they have often taken little account of such reading in their thinking. In particular, they have often ignored the very clear acknowledgement Freud made to his predecessors: philosophers, clinicians and novelists. Elsewhere, I have coined the term "the psychologist in literature"(McKellar, 1968, 1979). A better phrase would perhaps be the "psychologist-character" in literature. Some writers have created, in novels, short stories, or plays, a character through whom they have expressed their own psychological insights. Ibsen, for example, did this a great deal. It seems to have been these aspects of literature that interested Freud. Ellenberger refers to the "unmasking tendency" as a tradition in literature and psychology. By this he means the uncovering, or unveiling, of the destructive, self-punitive, and otherwise antisocial aspects of motivation at which human beings excel. Consider Ibsen. Again and again in an Ibsen play a "psychologist" emerges, an invented character who unmasks in the other characters self-deceptive and antisocial motives. The great exponents of the unmasking tradition such as La Rochefoucauld, Dostoevsky, Nietzsche, Ibsen, Freud and Anna Freud had a point of substance to make. Unconscious motivation is often revealed by a discrepancy between stated intentions and the obvious consequences of actions. La Rochefoucauld (1931) expresses this in an interesting overstatement in one of his *Maxims*: "if we judge love by its usual effects it resembles hatred rather than friendship." Von Hartmann also made his contribution to this tradition, with a cynicism and pessimism worthy of his much-admired Schopenhauer. He analyzed the various stages of development of human societies. First is the "childhood" phase: seeking happiness in the here and now. (Compare Frued's Pleasure Principle.) Next comes "youth": preoccupation with illusions about transcendental things, and perhaps with an afterlife or some other world. Then there is "manhood": the utilitarian concern with improving the welfare of people in present and future generations and, along with it, an overvaluation of education as a way of achieving this. Finally, there is "old age" (something like maturity, or perhaps the wisdom that allegedly comes from a Jungian analysis). This is described as a philosophical casting off of illusions, and in their place a combination of rational thinking and aesthetic contemplation. In this final phase we can see the influences of Schopenhauer, and of the Hindu and Buddhist adjustments to living which he sought - without the crudities of Hegelian influence - to bring into Western thinking. Pythagoras and Epicurus and Plato might well have endorsed this final stage of insight.

V. SEVEN PSYCHOLOGIES OF THE UNCONSCIOUS

The notion of the unconscious has some resemblances to a projective test, a Rorschach ink blot or a TAT picture awaiting interpretation. Before and after Freud, different psychologists and philosophers have given emphasis to one or another of these aspects. In Gestalt terms - my reference is to Wertheimer, Koffka and Kohler, not to Perls - what has been significant "figure" for one thinker has been passed over as less significant "ground" by others. The seven traditions enumerated are not, of course, mutually exclusive: a given thinker has, on occasion, given emphasis to more than one of them.

First is the dipsychic model of the personality: an emphasis upon two systems with imagery and other content erupting into consciousness as though from "outside." An elaborate development of this into a polypsychic model occurs in Jung: with a conscious comprised of numerous emotional systems or complexes, and an unconscious replete with many archetypes (e.g., Jung, 1940). A *second* tradition of thought placed emphasis on the unconscious as a seat of motivational forces, a fund of psychic energy. This is to be seen in Schopenhauer, Von Hartmann, Groddeck, as in Freud himself with the concept of the Id. The strongly pessimistic tradition of pessimism of Schopenhauer is prominent in the later Freud. Jung by contrast viewed the unconscious, and its energy, in a basically optimistic way: on occasion almost as an internalized helpmate and ally who advised and warned. A *third* tradition is developmentally oriented, and in the elaboration of this within his own system as the libido development theory Freud acknowledged his indebtedness to his close associate, Karl Abraham. Perhaps most of all it was Erik Erikson who has since accorded the developmental process the status of "figure" within his own system. Also to be noted is the elaborate developmental psychology of the child analysts, notably Melanie Klein. This has deviated considerably from the orthodox Freudian tradition. *Fourth* are depth psychologists such as Harry Stack Sullivan (1949) who were less interested in "an unconscious" than in the unconscious factors which complicate interpersonal relationships. The notions of transference and counter-transference in Freud seem to have been the predecessors of this reorientation of depth psychology. *Fifth*, emphasis has already been placed on the uncovering or unmasking tradition, prominent in Nietzsche and his literary predecessor, Dostoevsky. This, we have seen, greatly influenced literature and is of course of major importance in the ego psychology of Anna Freud (1936) and others. Analysis of the personality in terms of its preferred mechanisms of ego defense has been a focal point of this tradition of depth psychology. *Sixth*, some, like Freud himself, have placed their emphasis on processes of internalization: introjection of aspects of the external world into the personality. The superego is the product of such an introjection. Freud stressed the relation between warring internalized systems, especially in depressive psychosis (melancholia), on the one hand, and compulsive-obsessive neuroses on the other. Intra-psychic conflict has been a major point of emphasis of psychoanalysis. One way in which it can be handled is by the mechanism of compartmentalization, or "isolation," as Freud termed it. Here we come to an aspect of psychoanalytic theory that points to a polypsychic rather than dipsychic view of the psyche.

The *seventh* tradition relates to motivation and what I have often thought to be one of the fundamental points of emphasis of Freud. It is the notion of overdetermination and, relatedly, of multiple motivation. Behavior results from not one but a number of motives. The motives of a rational and ethically respectable kind most prominent to introspection may not be the most important. Moreover, there may be a surplus of motives: they may be more than sufficient to determine the behavior in question. In practice this means that when we have detected the obvious aspects of motivation - of the self or others - we go on looking for more. In these ideas Freud had his predecessors in, for example, Nietzsche, who emphasized how rationalizations may mask the less admirable aspects of motivation. We find them also in La Rochefoucauld, who has often been misinterpreted as defending a purely egoistic theory of motivation. His *Maxims* incorporate a much more subtle point: that along with other motivation we are likely *also* to find self-interest. Similarly, Freud's theory did not involve pure hedonism: the pleasure principle was but one of a number of motivational principles. As Flugel (1945) later put it: man is also, in certain respects, a pain-seeking animal. In this area Freud's important predecessors again included Schopenhauer, with his emphasis on powerful irrational motivational forces: and also his successors, such as, Flugel among the psychoanalysts. Other successors, such as those who use "transactional analysis" (e.g., Harris, 1973), claim that incorporated in the personality is the threefold motivation, that of "child," "an adult," and "a parent figure." This formulation relates to point six above, concerning internalization. Indeed, the seventh tradition under discussion is fundamental in that it is superordinate to all of the other six. By recognizing multiple motivation and overdetermination we are led to a dipsychic (or polypsychic) standpoint; to recognizing also the psychic energy aspects of motivation; the developmental aspects; the complications that enter into personal relations; the unmasking aspects; and the sixth point of emphasis just mentioned.

Seven traditional aspects of "unconscious" mental life have been considered. They were there for Freud to inherit and build upon. Philosophers, clinicians, mythologists and novelists all made their contributions. Of interest is the deep strain of pessimism represented by Schopenhauer which found its echo also in Freud. Schopenhauer himself made much of the concept of the "Veil of Maya": the wall of illusion that hides truth. To penetrate this "Veil of Maya" is to achieve something like release from the chains of the prisoners in Plato's famous Metaphor of the Cave, of *The Republic*. We look from the shadows to the realities which Plato found in his "Forms" and which Jung perhaps found in his archetypes.

This dichotomy between appearance and "reality" may be viewed in two ways: the metaphysical and the scientific. There is a great deal of the metaphysical in the writings of Carl Jung. Moreover, various religious systems, seemingly endless in their variety, have provided their chosen techniques like meditation. Freud, by contrast, claimed for his system the standing of a branch of science. As we have seen he was a major model builder, and as theorist made substantial use of analogies - drawn from mythology and literature - in building up his system. He was concerned with constructs that seemed to him to assist understanding of natural phenomena. In the tradition of Kant, many scientists would view science itself as a set of procedures that enable us to relate concrete phenomena to abstractions. Often these abstractions have become mathematical, though in the early stages they tend to be verbal. These assist communication, understanding and, ultimately, prediction. Apart from the metaphysical and the scientific at the physiological level, we have noted the illusions people have about the realities of their own personalities. The Freudian mechanisms of ego defense have provided a nomenclature - at the very least a system of labels - that has, on the whole, been acceptable to many orthodox psychologists. Along with repression itself we recognize denial, projection, and others that may help in the understanding of upsurges of imagery that seem to come from "outside," and impulses that seem to have origins outside the ego system. Important among them, I suspect, is the mechanism of compartmentalization or isolation. Consider the question again: "Where do these autonomous images and other upsurges that seem strange and alien to the personality come from?" (McKellar, 1977). Some have sought explanations that involve other than naturalistic interpretations. With these in mind also I turn to those whose formulations are polypsychic rather than dipsychic. Among these are systems which emphasize "subconscious mental life," "subsystems of the personality," and "dissociaton."

VI. POSSESSION AND DISSOCIATION

If the notions of dissociation and multiple personality are unfashionable in modern scientific psychology, that of spirit possession is even more so. But belief systems vary, as the work of anthropologists testifies. Even in this twentieth century, in Africa, Asia, South America, Afro-American, and in parts of Europe and North America, belief in "possession" remains widespread. A cross - cultural as well as a cross-temporal perspective is appropriate. In a study of 400 nonWestern cultures Erika Bourguignon (1973) found some institutionally-supported form of dissociation in 356 of them: in 89% of those studied. Earlier in history Anton Mesmer, when seeking to introduce what was after all a naturalistic construct - "animal magnetism," came into conflict with the exorcists. They defended *their* terminology. By contrast, others who have preferred the naturalistic to the supernatural have invoked the alternative construct, "dissociation." An example was William McDougall (1933), who interpreted both hypnotic states and the "trance personalities" of the seance room in dissociation terms. In fact, many of those who have sought a scientific explanation of "supernatural" phenomena have resorted to the same concept. A recent example is Wilson (1981), who has invoked dissociation and multiple personality to explain recent alleged cases of "reincarnation." Apart from noting differences of belief systems - naturalistic or otherwise - something additional may be noted about dissociation on the one hand, and possession on the other.

Ellenberger (1970) points to a difference between two kinds of "possession": somnambulistic and lucid. In *somnambulistic possession* the personality is wholly taken over by the possessing spirit: there is loss of awareness of identity, and subsequent amnesia for the episode. By contrast, in *lucid possession* personal identity is retained: the personality is aware of the "invading" spirit entity. I take an illustration of somnambulistic possession from one major investigator of Afro-American culture. In his classic study of Voodoo in Haiti, Metraux (1972) describes a girl, ordinarily a shy and retiring person becoming possessed by the "loa" or spirit Ezulie. For a time Ezulie, personifying jealousy, sexual love and eloquence, replaces the normal self. The girl screams, the screams subside, then she begins to dance, gyrating slowly, her movements provocative and sensuous. The period of possession comes to begin an end, and the next day there is amnesia for the whole episode. Voodoo possession seems to be predominantly of this somnambulistic kind. Its social function has been examined by Metraux himself, and subsequent investigators (e.g., Brown, 1976). There is a large parthenon of loa available to assume possession. Poverty - stricken peasants, in one of the poorest countries of the world, can choose and can, for a time, escape into their fantasy. While "possessed," they become objects of fear, reverence and awe. Adults can engage in actions of their choice with this supernatural excuse. Children can, in a way, educate their parents: while possessed are listened to with attention and respect. As Karen Brown (1976) puts it "within a single person there are many selves . . . in Haiti these manifest themselves through the various personalities of the gods" (p. 289).

The explanation offered is a polypsychic one, of personality subsystems. Moreover, the anthropological distinction between such somnambulistic possession and the lucid kind has an interesting parallel in case histories involving dissociation. In their survey of multiple personality cases Taylor and Martin (1944) distinguish between alternating personality and co-consciousness. In alternating personality, something similar to "somnambulistic possession" seems to occur: a total taking over, loss of identity and amnesia. The classical case of Rev. Ansel Bourne is a good example (William James, 1890). A parallel to the alternative kind, lucid possession is to be found in some of the other famous cases. For example, in Morton Prince's patient, Miss Beauchamp, the "Sally" subpersonality could co-consciously overlook the other systems. She could, for example, impose her own imagery upon them while they slept and thus produce what they - on waking - labeled "dreams."

Naturalistic language involving the notion of subsystems provides a way of formulating how "possession" can function. Other seemingly supernatural occurrences may be considered from a similar standpoint. Elsewhere (1979) I have suggested that perhaps certain poltergeist phenomena could be explained in this way. Morton Prince (1905) retained a naturalistic explanation of multiple personality in the case of Miss Beauchamp. Yet, the torments which the Sally personality inflicted on the other Miss Beauchamp' at times came close to what others might have accepted as the activities of a malevolent poltergeist. Writing of such phenomena Laird (1917) argued that multiple personality is not new: it has been there for a long time, differently labeled in "demonic possession," "lycantropy," and the prophesies of the oracles of ancient Greece.

Dissociation involves systems of ideas with emotional and motivational accompaniments split off from the main personality, but continuing to function. Sometimes, as in lucid possession, they are re-interpreted as an invading entity; on occasion, as in somnambulistic possession they assume control of the total organism, and - for a time - a new identity. Using this formulation we may speak of dissociation not as an all-or-nothing phenomenon, but of degrees of dissociation.

VII. THE POLYPSPYCHIC PERSONALITY

Ellenberger finds the origin of the term "polypsychism" in the magnetizer "Durand (deGros)" (Ellenberger, 1970, p.146). He underlines the importance of the subsequent work of Pierre Janet, and the seemingly unwise neglect of the dissociationist tradition. The success of the psychoanalysts in establishing their alternative language, he suggests, partly explains this neglect. Janet's ideas became known in the United States, where he had visited and lectured, and particularly the Boston area. They appealed to William James (1890, Vol.1) of Harvard, who refers to "splitting up of the mind into different consciousnesses" and who adds that Janet "designates by numbers the different personalities which the subject may display" (p. 210). The metaphor of "splitting" came very much into use among other Boston investigators, Boris Sidis and Morton Prince. Outside mainstream clinical psychology, investigators became interested in the paranormal. Those seeking to understand not only "possession," but also crystal gazing and automatic writing and the behavior of spiritualistic mediums often resorted to the language of dissociation. Among recent contributions to this area may be noted the work of Zusne and Jones (1982). As Zusne elsewhere (1983) puts it, "mental imagery, dissociated states of personality, and magic in its multifarious forms . . . from automatic writing to voodoo possession are inextricably linked. There is an enormous field here, waiting for someone to work out the relationships" (p. 28). In an earlier period William McDougall (1926) deplored the absence of "dissociation" from psychoanalysis: his own abnormal psychology text also paid considerable attention to the paranormal. It leaned heavily on the concepts of Janet and his vigorous supporters among the Boston psychologists.

Among those interested in dissociation, and the polypsychic view of personality, two somewhat different traditions may be detected. Pierre Janet himself was mainly interested in automatons, hysteria, subsconscious perception in the neuroses, and hypnotism. Morton Prince - like McDougall after him - used dissociationist thinking in an approach to personality psychology more generally. It may be noted also that some of Janet's original distinctions have, perhaps unfortunately, been lost. He certainly did not identify somnambulism with sleep walking, and indeed had much to say about diurnal somnambulisms. Unlike some later investigators and textbook writers, he differentiated these diurnal somnambulisms from fugues. Somnambulisms are of brief duration, and are responsible for bizarre and attention-arousing behavior. Fugues, for Janet, differed not only in their longer duration. He observed that the person concerned would "behave normally," without attracting attention, even unobstrusively. He writes, "they take railway tickets, they dine and sleep in hotels, they speak to a great number of people . . . they are not recognized as mad" (Janet, 1907, p. 60; McKellar 1979, pp. 18 ff and Glossary of Terms) Similarities between these two kinds of automatisms should not blind us to noticing important and far from subtle differences between them.

Multiple personality and the schizophrenias have been endlessly confused by popular thinking and the media under the rubric of "split mind." Often the same confusions have been found in the proceedings of courts of law. Despite this, what might be called the "Jekyll-Hyde syndrome" has figured prominently in imaginative fiction. This has included the literature of France, Germany, Scotland, England and the United States. We may notice also the film, for example, Alfred Hitchcock's classic *Psycho*. In this - as in Edgar Allen Poe's story, "William Wilson" - we see an instance of what Janet would have labeled a "dominant somnambulism," a takeover of the primary personality by a stronger, more robust, secondary personality. Such a happening also forms the plot of the short story by Dostoevsky, "The Double." In the Jekyll - Hyde story also it will be noted Dr. Jekyll reported how the alien personality would intrude by "scrawling in my own hand blasphemies on the pages of my books" (McKellar, 1979, p. 31). Later, this automatic writing gave way to something resembling somnambulistic possession: total control of the whole organism by Mr. Hyde. Similar things seem to have happened in the Miss Beauchamp case. Morton Prince was familiar with Stevenson's writings. In a letter to me, Prince's successor at the Harvard Clinic, Professor Henry Murray reported that Prince would often talk to him of "Dr. J. and Mr. H" (personal communication, 1978). Undoubtedly the insights of men of literature have, in many different ways, influenced the formulations of professional psychologists, about dissociation phenomena.

Eugene Bleuler in Zurich and Janet in France were interested in and exposed to different clinical cases. Justification may be found for maintaining a distinction between the "molecular" dissociations of the shattered schizophrenic personality and the molar dissociations of hysteria and multiple personality. The DSM III and its predecessors maintain such a distinction. Yet despite obvious differences between the two, some similarities can be detected. Consider the original Breuer-Freud *Studies on Hysteria*. Highly noticeable in these cases is the frequency with which "hearing voices" and other hallucinations occurred with these "hysteria" patients. To conceptualize this fact in terms of different personality "systems" seems to have had some appeal to Breuer. Moreover, Freud himself later wrote of processes that "enjoy a large degree of mental independence, as though they had no communication with one another." He goes on to write, in a speculative vein of assuming "the existence not only of a second consciousness, but a third, fourth, perhaps an unlimited number of states of consciousness, all unknown to us and one another" (Freud, 1915/1957, p. 170). The question arises of whether some kind of reconciliation, or at least translation of concepts, can be achieved between psychoanalytic and alternative models of abnormal mental life. My emphasis will be on the relations of "the unconscious" to "dissociation."

VIII. THE UNCONSCIOUS VERSUS DISSOCIATION

A frontal attack on the task of translation was made some years ago by the English psychiatrist, Bernard Hart (1939). Hart was interested in repression and the unconscious, on the one hand, and dissociation and subconscious mental states on the other. His conclusion was that the terminology of dissociation is largely descriptive: It provides labels for certain phenomena of psychopathology. The evidence for fugues, somnambulisms, trance personalities and the like is parallel to the evidence for mental life in general. Hart also found the dissociation terminology appropriate to the phenomena of hypnotic states; he had, however, reservations about the "splitting" metaphor in relation to consciousness. By contrast, the alternative psychoanalytic terminology of repression and the unconscious was, he argued, explanatory rather than descriptive.

Repression he viewed as "an imagined entity created in order to explain phenomenological facts." We may, he added, deduce consequences from it and thus attempt to verify "the validity of the conceptual construction" (Hart, 1939, p. 166).

Since Hart's time there has been much attention to the different types of "conceptual constructs." In learning theory, for instance, it was once almost obligatory to cite the distinction between hypothetical constructs and intervening variables (MacCorquodale & Meehl, 1948). Hart's distinction between labels for phenomena and concepts designed to explain them seems fundamental, and is perhaps less dated. Both are familiar in the other sciences. The Linnaean system of labels within zoology and botany specifies in a descriptive way. By contrast, the Darwinian concept of natural selection as one of the mechanisms of evolution is explanatory. Despite the fact that Linnaeus himself rejected this, his own theoretically neutral labels - and their elaboration - survived. Within psychology we enounter fugues, somnambulisms, amnesias and - more rarely - multiple personality. The term "dissociation" is a superordinate name for such phenomena. It has survived in recent times into, for example, the DSM III, the American Psychiatric Association official classifactory system. Rarity or otherwise of psychiatric syndromes itself is something requiring caution; likewise, the implication of pathology and maladjustment. Recently the psychologist Colleen Ward (1983) has made a plea that "in the overall analysis of imagery, magic and mental health, the phenomena should be appraised in a cross-cultural context." She takes as an illustration a contemporary group in Trinidad and Tobago and suggests that some such dissociationist practices may be therapeutic: "we should not approach them with preconceived bias and underlying assumptions of intrinsic abnormality" (p. 20). Earlier I have likewise indicated that to a quite considerable extent voodoo "possession" in Haiti may well perform important social, even therapeutic, functions. Leonard Zusne has likewise defended a broader attitude to phenomena of dissociation in the areas frequently assessed as paranormal. He assesses "multifarious forms of anomalistic psychological phenomena - from automatic writing to voodoo possession" as being linked with "dissociated states of personality" (Zusne, 1983, p. 20). Those who have sought to understand, within the scientific framework, cultures other than our own, with their overtones of the seemingly paranormal, continue to find the concept of dissociation useful as a label. Hart's defense of such labeling of phenomena seems to have withstood the passage of time. Some societies seem to positively encourage dissociation, and even phenomena remarkably similar to or identical with multiple personality. Labels in this area continue to be needed. But how does dissociation differ from repression?

In a paper earlier than Hart's, Kardiner - a cross-cultural psychologist - identified dissociation not with repression but with something related to it. He wrote, "one can use the word dissociated to designate *return of the repressed*" (Italics mine). Kardiner added that "what appears to be detached from the aims and purposes of the individual is still a living part of him" (Cited by Hart, 1939, p. 175). Of interest is a very similar statement by Freud himself, who wrote, "the memory trace of the repressed idea has, after all, not been dissolved; *from now on it forms the nucleus of a second* psychological group" (Freud, 1894/1962, p. 49). In the same paper Freud likened such a group to "a sort of parasite." Parasites function; they are not inert. A different metaphor, but carrying a similar meaning, is used by Karl Menninger in his attempt to explain the idea of the Freudian unconscious. He likens the mind to a theater, with consciousness as the stage. The stage manager, the superego, exerts control over entry to this stage. But, as Menninger (1946) puts it, referring to repressed ideas and impulses, "*all* the actors . . . want to act" (p. 279). They are vigorous in their efforts to get on the stage, this is, to enter consciousness. Mowrer (1950) has used yet another analogy of a "medieval city" outside whose walls certain disruptive elements - under conditions of repression - are banished. They clamor for readmission: Mowrer's analogy for threatened return of the repressed. Many different formulations, and analogies have been used to underline this aspect of repression. To turn from analogy and metaphor to definitions, the *Critical Dictionary of Psychoanalysis* defines "return of the repressed" as "the involuntary irruption into consciousness of unacceptable derivatives of the primary impulse, not the dissolution of the primary repression" (Rycroft, (1968, p.142). The upsurges of imagery labeled "hallucinatory voices" that troubled Miss Beauchamp, Eve and other classic multiple personality cases are well described in the analogies chosen by Menninger and Mowrer. They resembled unwanted actors intruding on the stage, or disruptive rebels that had scaled the walls and reentered Mowrer's walled city. Although the concept of repression does not figure in his work, on the therapy side Mardi Horowitz (1983) makes considerable use of "unbidden images." These he assesses in a way very similar to the earlier writers cited: "Unbidden images are, first, and foremost a failure in repression" (p. 180). He points to a wide range of emotional responses in such imagery, ranging from surprise to fear and panic. Such imagery, like hallucination, is experienced as coming from "outside" though its origins lie within the mental life of the imager himself. Some, as we have seen, refer to it as a "return of the repressed" or, like Horowitz, as a "failure of repression." An alternative terminology labels it as a functional dissociated system, invading conscious awareness. Another puzzling phenomenon of interest is deja-vu. A number of theorists have sought to explain this also in terms of failure of repression (Neppe, 1983). The alternative synonym for this - dissociation - may again be noted.

The important thing about dissociation seems to be that it is in some way functional. As Sidis and Goodhart (1905) using dissociation rather than the Freudian terminology, put it, systems "that have become dissociated carry on their functioning activity side by side with the main systems." Hart prefers what he calls a "functional" conception of dissociation. He is critical of the spatial analogy of "splitting"; to him it is more like what happens when we change gears in a motor vehicle. Hart (1939) writes: "it is an affair of gearing the various elements of mental machinery being organized into different functioning systems by throwing in of the appropriate gear." In the realm of analogy many are available; for example moving the dials of a radio set and tuning in to a different radio station and a new program. Subsequent advances of technology provide still further analogies. For example, the small hand calculator I possess has different memory systems which operate when I press the appropriate key. The investigators of Eve were able to "call out" the personalities of Jane, Eve White, and Eve Black. At one American Psychological Association meeting, I saw a film of them doing just this with the original patient. Likewise, I can call on my calculator to tell me the time or the date; or I can instruct it to resort to its other systems by operating as a stopwatch or a calculator. The component systems do different things, and "amnesic barriers" exist between them when the calculator is functioning normally. In Freudian terminology the mechanism of compartmentalization or "isolation" remains effective, or as Hart puts it elsewhere (1936), we are very much dependent on the "logic tight compartments" of our own personality. In the Beauchamp case, B1, B1v and Sally coexisted in an arena of mutual antagonism. The compulsive neurotic characteristically declares "I don't know what makes me do it . . . I know it's irrational . . . I can't help it." Or in more everyday life situations, the would-be sleeper is frightened, surprised, amused, or even angered, by the eruption of hypnagogic imagery; only with some difficulty does he recognize his own authorship of it. Dissociation is a concept which describes a multitude of phenomena - normal, abnormal, and paranormal. Concepts like isolation, repression and return of the repressed may help in the understanding of its underlying dynamics.

IX. THE PROCESS OF THERAPY

Words may mislead us by their associations when considering the relation of "the unconscious" to therapy. Consider first dreams as "the royal road to the unconscious." The unconscious is not something we "sink into" when we sleep, and emerge from on waking. It is not like the "Never-Never land" portrayed by J.M. Barrie in *Peter Pan* which the island children visit in their dreams, and where anything can happen. The Freudian unconscious is not a place; its relation to dreams can be formulated in other terms. Sleep itself is, if anything, an altered mental state - not "unconsciousness" - and during it thinking and imaging, feeling and motivation, continue to go on. It is not surprising that the thinking of sleep - like the thinking of wakefulness - should reflect the ideas, emotions and motives of the personality. If such thinking, in the form of dream imagery, is recalled we have material for dream interpretation, such material providing Freud's "royal road" to understanding the person under study. During sleep the person has been "conscious" at least in the sense that he has been experiencing such imagery, sometimes stimulated by contemporary sense perception. In the interesting atypical case of lucid dreaming, while asleep he appears also to have been conscious of his own identity, and of some memories of the kind available to normal wakefulness. During therapy, recall of the imagery of sleep may be used as a means of eliciting information not otherwise consciously available. There are, as we have seen, alternative ways of tapping such information: hypnotism, projective testing, and - in some people - crystal gazing and automatic writing.

Freud's own recommended method of dream analysis may provide a reorientation to what happens in psychotherapy. The first thing needed to "interpret" a dream is the presence of the dreamer, and the eliciting from him of his free associations. It is not a matter of looking up a list of standard symbols and their meaning from some dreambook. Freud is both explicit, and emphatic, in rejecting this procedure. It must be admitted that he did, on occasion, lapse from his own sensible rule of free association, notably in his discussions of the symbolism of literature and paintings. Yet his stated emphasis on free association is unambiguous. It involves establishing relations between the elements of the dream and, through association, relating these to the personality. To resort to Spearman's terminology, what is involved is seeking to link things up, that is, to "educe" relations (e.g. Spearman, 1922). Much of this process seems to involve linking up imagery, memory and imagination images of the kind available also in wakefulness. The emphasis is on becoming aware of relations.

Spearman was not a therapist, nor do I claim to be. Moreover, he was writing about other things. His Noegentic Theory was an attempt to analyze the psychology of intelligent thinking. This he identified with the ability to establish relations. Similarities may be found between thinking intelligently and thinking straight about oneself, that is, having insight. "Intelligent" is an adjective which can be applied to the differing performance of the same individual in a variety of circumstances. When flustered, emotionally aroused, or under the influence of prejudice, an intelligent person can be noticeably unintelligent. In one of his lesser known works Freud (1907/1959) makes the same point. He refers to it as "an astonishing fact, and one that is too generally overlooked . . . people of even the most powerful intelligence react as though they were feebleminded" (p. 71). Applying to this Spearman's formulation this means they are, on occasion, deficient in their ability to apprehend experience, educe relations, and educe correlates (Spearman, 1922). Freudian therapy involves assisting them to relate things to one another better, when the barriers to doing this are emotional and motivational. Denial labels the situation when the obvious facts of experience are not apprehended, because of contrary emotion. Consider the Oedipus of Sophocles in *Oedipus Rex*. For a long time he was remarkably unable to apprehend the fact - obvious to others - that the woman he had married was his own mother. In real life Freud points out that there is a parallel. Many people resist recognizing the influence of their parent imago on their choice of a love object, or spouse. There is a parallel with educing relationships: in analysis a patient is made aware of the fact that a person whom he strongly dislikes is an uncomfortable reminder to the patient of aspects of himself he has rejected. Relational thinking is improved by therapy, and in the direction of thinking more intelligently. When emotion is involved there can be barriers to grasping relations. Freud speaks of "resistances." It may be logically deplorable to hold incompatible opinions but, as Hart shows, it is psychologically commonplace. Therapy seeks to minimize this, to encourage a person to think straight about himself and his interpersonal dealings with other people. Much of the therapy process involves something resembling putting aside emotion and prejudice in a way that otherwise intelligent people do not always do when thinking about issues like politics, religion, or the motivation of those they dislike. Much that is taking place in therapy is often not a matter of "excavating" unconscious memories from areas of depth and darkness. It is the linkage between ideas, and often between subsystems of the personality, that is unconscious. Writing as a psychoanalyst, in one place Ernst Kris comes close to this terminology and orientation to therapy. He writes that insight is achieved "by re-establishing links that have been lost" (Kris, 1950, p. 480). This statement by Kris acknowledges that psychoanalysis is concerned with establishing intrapsychic communication through something like the "walls" of Hart's "logic-tight compartments." Can we view psychoanalysis as a technique for ridding the personality of some of the undesirable aspects of "dissociation"? A careful reading of the case history of Breuer's famous patient suggests that Anna O - and perhaps Breuer himself - might have agreed. As Ahsen (1972) has argued, the patient in this case may have made substantial contributions of her own to the development of what was to become psychoanalysis. An examination of the case history of Morton Prince's Miss Beauchamp, along with Anna's, suggests the further possibility that the two young women may well have been very similar personalities. Their neuroses and related problems of adjustment were conceptualized differently by their respective therapists. As controversies developed, and as Freud's - as opposed to Breuer's - influence over theory increased, these differences became greater and greater. Had Miss Christine Beauchamp come to Morton Prince rather than Freud's collaborator, she would certainly have been a case of "multiple personality." Moreover, the account of the case by Breuer himself differs but little from the terminology used by Janet. To quote from an earlier line of the poem, the Kasidah,

When doctors differ, who decides . . .

For my part I would suggest a case conference during which they might find their differences of diagnosis possible to resolve.

X. CONCLUSIONS

In conclusion, an analogy of my own. We can map human personality in a variety of ways. Two historically important maps have emerged from study of the phenomena of unbidden imagery and other autonomous processes. Reconciling the two maps, combining the information contained in each, may now be possible. Failing such an achievement, perhaps we should at least be concerned with establishing a "bridgehead" for those prepared to defend the banner of "translation." Under this it may be possible to advance and take over new territory from the realm of the unknown, alias the prescientific and the allegedly supernatural.

This chapter has concerned the psychology of the unconscious, theories about it, and their historical roots. Psychology, like other sciences, is a system of constructs which are devices to assist the understanding of natural phenomena. The constructs of different theorists reflect the anologies which underly them. Theory involves a process of abstracting a system of relations from one subject matter and applying it to another. In theories about "unconscious" mental life - as about imagery - a wide range of different constructs has emerged through time. This culminated in Freud, who was an influential model builder. In developing his system he made substantial use of analogies and metaphors. Other theorists today, for example in the imagery field, continue to do likewise. A promising development of recent years has been an emerging interest in the place of analogy in human thinking. This chapter has discussed the intellectual heritage on which Freud built and from this standpoint. Important among his predecessors were Schopenhauer and Von Hartmann. From these thinkers and others, including Nietzsche, may be noted seven traditions each representing a major point of emphasis about the unconscious.

Moreover, two major models of the personality have been influential. Both seem highly relevant to understanding "unbidden imagery" and other upsurges of autonomous mental life - surprising, frightening, even anger-provoking - from outside the ego system. These two models are the dipsychic (conscious versus unconscious), and the polypsychic (many subsystems of the personality). In comparing them we may ask: To what extent are the notions of "dissociation" and "the unconscious" synonymous? This question is explored. It is concluded that both traditions serve to highlight different aspects of mental life, but that some concern with translation between them is now possible.

Two of the most influential case histories - Anna O and Christine Beauchamp - in the history of psychology invite reconsideration. Were they very similar personalities? Yet, very different constructs, influenced by alternative choices of analogy, stemmed from the study of these two people. A challenge to say it in different words, to reformulate one's preferred constructs, often results in better comprehension. Such translation assists removal of mere semantic differences. In the words of the poet cited, two seemingly opposed theories may both be right and/or both may be wrong. In either case interest in the analogies that theorists use and more concern with the activity of translation offers promise for the future.

REFERENCES

Adler, A. (1932). *The practice and theory of individual psychology*. London: Kegan Paul.
Ahsen, A. (1972). Anna O: Patient or therapist? In V. Franks & V. Burtle (Eds.), *Women in therapy*. New York: Brunner/Mazel.
Block, N. (Ed.). (1981). *Imagery*. Cambridge, MA: MIT.

Bourgignon, E. (Ed.). (1973). *Religion, altered states of consciousness and social change.* OH: Ohio State University Press.

Breuer, J., & Freud, S. (1955). *Studies on hysteria.* London: Hogarth Press. (Originally published, 1893-1895)

Brown, K.M. (1976). *The veve of Haitian voudou: A structural analysis of visual imagery.* Unpublished doctoral dissertation, Temple University, Philadelphia.

Buckler, W.E., & Sklare, A.B. (1966). *Essentials of rhetoric.* New York: Macmillan.

Dennett, D.C. (1981). Two approaches to mental images. In N. Block (Ed.) *Imagery.* Cambridge, MA: MIT.

Ellenberger, H.F. (1970). *The discovery of the unconscious.* New York: Basic Books.

Flugel, J.C. (1945). *Man, morals and society.* London: Duckworth.

Foster, M.A. (1921). *Studies in dreams.* London: Macmillan.

Freud, A. (1936). *The ego and the mechanisms of defence.* London: Hogarth.

Freud, S. (1962). The neuro-pychoses of defense. In J. Strachey (Ed. and Trans.), *Collected works* (Vol. 3). London: Hogarth Press. (Original work published 1894)

Freud, S. (1933). The interpretation of dreams. In J. Strachey (Ed. and Trans.), *Collected works* (Vol. 4). London: Hogarth Press. (Original work published 1900)

Freud, S. (1959). Delusions and dreams in Jensen's Gradiva. In J. Strachey (Ed. and Trans.), *The standard edition of the complete psychological works of Sigmund Freud* (Vol. 9). London: Hogarth. (Original work published 1907)

Freud, S. (1957). The unconscious. In J. Strachey (Ed. and Trans.), *Collected works* (Vol. 14). London: Hogarth Press. (Original work published 1915)

Freud, S. (1955). The ego and the id. In J. Strachey (Ed. and Trans.), *Collected works* (Vol. 19). London: Hogarth Press. (Original work published 1923)

Galton, F. (1883). *Inquiries into human faculty.* London: Macmillan.

Gordon, R. (1962). *Stereotypy of imagery and belief as an ego defence.* London: Cambridge University Press.

Groddeck, G. (1928). *The book of the it.* New York: Nervous and Mental Diseases Publishing Co.

Harris, T.A. (1973). *I'm ok - you're ok.* London: Pan Books.

Hart, B. (1936). *The psychology of insanity.* London: Cambridge University Press.

Hart, B. (1939). *Psychopathology: Its development and its place in medicine.* (Esp. pp. 154-175) London: Cambridge University Press.

Hartmann, E. von. (1931). *Philosophy of the unconscious.* London: Kegan Paul.

Horowitz, M. (1983). *Image formation and psychotherapy.* (Rev. ed. of *Image formation and cognition*) New York: Aronson, 1983.

Jaensch, E.R. (1930). *Eidetic imagery.* London: Kegan Paul.

James, W. (1890). *The principles of psychology, Vol. 1 and 2.* London: Macmillan.

Janet, P. (1907). *The major symptoms of hysteria.* (2nd ed.) New York: Macmillan.

Jaynes, J. (1976). *The origins of consciousness in the breakdown of the bicameral mind.* Boston: Houghton Mifflin.

Jung, C.G. (1940). *The integration of the personality.* London: Kegan Paul.

Kris, E. (1950). On preconscious mental processes. *Psychoanalytic Quarterly, 19,* 540-560.

Laird, J. (1917). *Problems of the self.* London: Macmillan.

Leaning, F.E. (1925). An introductory study of hypnagogic phenomena. *Proceedings of the Society for Psychical Research, 35.*

MacCorquodale, K., & Meehl, P.E. (1948). On the distinction between hypothetical constructs and intervening variables. *Psychological review, 55,* 95-107.

McDougall, W. (1933). *An outline of abnormal psychology.* London: Macmillan.

McKellar, P. (1967). *Imagination and thinking.* New York: Basic Books. (Original work published 1957)

McKellar, P. (1963). Three aspects of the psychology of originality in human thinking. *British Journal of Aesthetics, 3, 4,* 129-147.

McKellar, P. (1968). *Experience and behaviour*. Harmondsworth: Penguin.

McKellar, P. (1977). Autonomy, imagery and dissociation. *Journal of Mental Imagery, 1,* 93-107.

McKellar, P. (1979). *Mindsplit: The psychology of multiple personality and the dissociated self*. London: Dent.

McKellar, P., & Simpson, L. (1954). Between wakefulness and sleep: Hypnagogic imagery. *British Journal of Psychology, 45,* 266-276.

Marks, D.F. (1983). Mental imagery and consciousness: A theoretical review. In A.A. Sheikh (Ed.), *Imagery: Current theory, research and application*. New York: Wiley.

Marks, D.F., & McKellar, P. (1982a). The nature and function of eidetic imagery. *Journal of Mental Imagery, 6*(1), 1-28.

Marks, D.F., & McKellar, P. (1982b). Eidetic imagery de-reified. *Journal of Mental Imagery, 6*(1), 100-114.

Menninger, K.A. (1946). *The human mind*. New York, Knopf.

Metraux, A. (1972) *Voodoo in Haiti*. London: Deutsch, 1972.

Mowrer, O.H. (1950). *Learning theory and personality dynamics: Selected papers*. New York: Ronald.

Neppe, V.M. (1983). *The psychology of deja vu*. Johannesburg: Witwatersrand University Press.

Ortony, A. (Ed.) (1979). *Metaphor and thought*. London: Cambridge University Press.

Paivio, A. (1979). Psychological procession: The comprehension of metaphor. In A. Ortony (Ed.), *Metaphor and thought*. London: Cambridge University Press.

Pinker, S., & Kosslyn, S.M. (1983). Theories of mental imagery. In A.A. Sheikh (Ed.), *Imagery: Current theory, research and application*. New York: Wiley.

Prince, M.H. (1978). *The dissociation of a personality*. London: Oxford Unversity Press. (Original work published 1905)

Rochefoucauld, F. (Duc de la). (1931). *Maxims of La Rochefoucauld, posthumous text and translation*. (K. Platt, Ed.). London: Haworth.

Rycroft, C. (1968). *A critical dictionary of psychoanalysis*. Harmondsworth: Penguin Books.

Schopenhauer, A. (1883). *The world as will and idea*. London: Trubner. (Original work published 1818)

Schopenhauer, A. (1958). *The world as will and representation*. (E.F.J. Payne, Trans.). New York: Dover. (Original work published 1918)

Sidis, B., & Goodhart, S.P. (1905). *Multiple personality*. New York: Appleton.

Spearman, C. (1922). *The nature of intelligence and the principles of cognition*. London: Macmillan.

Spearman, C. (1930). *Creative Mind*. London: Nisbet.

Sullivan, H.S. (1949). *Conceptions of modern psychiatry*. Washington: William Alanson White Foundation.

Taylor, W.S., & Martin, W.F. (1944). Multiple personality. *Journal of Abnormal and Social Psychology, 39,* 281-201.

Ward, C. (1983). Imagery, dissociation and mental health: A cross cultural perspective. *International Imagery Bulletin, 1,* 19-20.

Wilson, I. (1981). *Mind out of time: Reincarnation claims investigated*. London: Gollanz.

Zusne, L., & Jones, H.W. (1982) *Anomalistic psychology: A study of extraordinary phenomena of behavior and experience*. Hillsdale, NJ: Lawrence Erlbaum.

Zusne, L. (1983). Imagery as magic, magic as natural science. *International Imagery Bulletin, 1,* 28-29.

CHAPTER 3

On Imaginal Mental Activity and the Imaginal Mental State

Gosaku Naruse

I. CONDITIONED IMAGERY

1. The Appearance of Conditioned Imagery

This chapter will begin with a brief description of a series of experiments on conditioned imagery in the posthypnotic hallucinatory state which were carried out from 1950 to 1960 by the author and his colleague, Torao Obonai. The basic procedure of the experiments was divided into two parts: (1) registration of perceptual experiences under hypnotic trance, and (2) observation of imagery in the waking state with posthypnotic amnesia of the preceding registration.

In the registration period, two kinds of perceptual stimuli were given as a pair; for instance, a figure drawn on a card as a visual stimulus and a sound of bell or buzzer as an auditory one were presented to the subject several times at the same time or in succession. The purpose of such a procedure is to condition a visual experience to the sound, and/or to make an association-bond between two memory traces of the two kinds of perceptual experience as shown in Figure 1A.

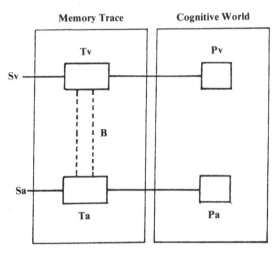

Sv: visual stimulus
Pv: visual perception
Sa: auditory stimulus
Pa: auditory perception
Tv: memory trace of Pv
Ta: memory trace of Pa
 B: audio-visual combination bond in memory trace

Figure 1A. Audio-visual combination by bond at the registration period

After the subject was awakened from hypnosis with posthypnotic amnesia for his whole experience of the figure-sound conditioning, the procedure of imagery observation was carried out. Asking the subject to report any visual experience which he "saw" on a blank card on which there was no figural stimulus, the experimenter presented again the sound of the bell or buzzer which was one-half of the stimulus pair at the registration (see Figure 6A). Most of the hypnotic subjects reported some visual experiences of imagery or hallucination without identifying them with their visual presentation in the preceding registration in the hypnotic trance.

Such imagery had been named "conditioned sensations" by Leuba (1940) and as "imagery in the posthypnotic hallucinatory state" by Naruse and Obonai (1953-1956), and Naruse (1962b) called it *conditioned imagery*. These authors confirmed that such imagery was able to appear not only in visual experience but also in the auditory, tactile and olfactory modalities. For the purpose of detailed examination of image modification, only visual imagery is described in this paper.

The images appeared in various degrees of clearness; some of them were classified as "hallucination," others were more like a memory image, idea or representation, and others were in between the two. It seemed that the deeper the preceding trance, the clearer and the more vivid was the imagery, and the more correct was the form. By correctness the author means a lack of distortion or variation in the imagery compared with the originally presented figure. Distortions of imagery were liable to occur along the space axis in location (displacement), in shape (enlargement or contraction), through decompositions of the figural components, or modifications according to the meaning of the original figure and the subject's attitude or mental set.

There were two types of imagery appearance of complex or meaningful figures. One was a clear and correct image which the subject failed to guess as a whole or identify. The other was a vague and incorrect one which the subject was able to grasp as a whole or identify. The former type was apt to appear in subjects in deeper hypnosis and the latter in subjects in a lighter trance.

These kinds of imagery tend to disappear when subjects try to see or perceive what is on the card or locate the imagery correctly in their perceptual field, and also when subjects pay attention to the sound. These facts seem to imply that active attention to objective reality is a disturbing factor for imagery. And for most subjects the appearance of imagery was not at the same level as visual perception in cognitive experience. However, a few subjects were not able to differentiate between images and percepts . The above facts seem to show that for the appearance of imagery it is necessary to assume that the subject has a mental set toward imagery or unreality which blocks or inhibits some usual perceptual pathway in cognition. Figure 1B shows the pathway of a stimulus through a memory trace for visual imagery during imagery observation.

After experiencing two kinds of paired presentations of acoustic-visual combinations which were separately formed through the conditioning procedure under preceding hypnosis (for instance, a *bell sound with an x figure* and a *buzzer sound with a ⊙ figure*), the subject reported that he experienced double imagery or two overlapping images during the presentation of the two kinds of sound (bell and buzzer) at the same time. Corresponding to the relative intensity of the two sounds or their localization, the double imagery changed its appearance in clearness or construction.

Two components of double imagery not only coexisted with each other at

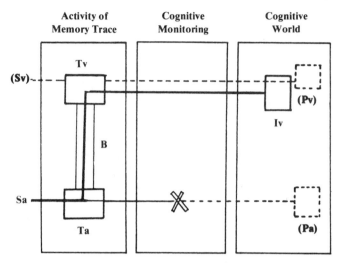

Iv: visual imagery evoked by Sa through Ta, B, Tv. Sometimes Iv ≒ Pv,
Iv = Pv or Iv ≠ Pv in subject's visual experience.
X : blocked pathway

Figure 1B. Association between auditory stimulus and visual imagery by blocking of paths of visual perception and auditory perception at the observation of imagery.

the same time but were also composed into a unitary or integrated image. A suitable degree of depth of the preceding hypnosis was needed for the occurrence of such a modification of the stimulus relationship, movement of the stimulus, or integrated imagery. When it was this deep, however, the imagery was too correct for modification or integration (Naruse & Obonai, 1955a, 1955b). Figure 2 shows the hypothesized mechanisms for the above-described experimental observations.

2. The Imaginal Image and the Perceptual Image

No subject reported any change of image size in the posthypnotic state corresponding to changes in the observation distance (distance between the observer and the card used for image projection). However, when the size of the card used for projection was changed, the size of the imagery changed. The imagery also altered its direction when an alteration of the direction of the card occurred. When color imagery was projected onto a card of the complementary color, the subject did not report any mixed color of imagery, but saw the same color as the original. And when two images, complementary in color to each other, were overlapped no subject reported any color mixture in imagery (Naruse & Obonai, 1953, 1955a). From the above-mentioned facts, it may be inferred that the imaginal field dissociates itself from the perceptual field through the activity of cognitive monitoring.

When one image was projected onto a card upon which the other figure

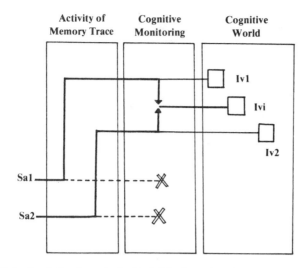

Sv1: visual stimulus paired with Sa1 in the registration period (RP)
Sv2: visual stimulus paired with Sa2 in RP
Sa1: auditory stimulus paired with Sv1 in RP
Sa2: auditory stimulus paired with Sv2 in RP
Iv1: visual imagery evoked by Sa1 in the imagery observation period (OP)
Iv2: visual imagery evoked by Sa2 in OP
Ivi: visual imagery composed into a new image made of Iv1 and Iv2
Iv1 and Iv2 coexist for deep-trance subject
Iv1 and Iv2 interact and integrate into one unitary image for the subjects
  who are in a suitable depth of trance

Figure 2. Mechanism of overlapping of two mental images.

was drawn as a perceptual stimulus, competition and compound interactions between the two images occurred. In cases of competition, a dominant perceptual image inhibited the appearance of inferior imagery, and dominant imagery inhibited the experience of perception in the cognitive field. There were two kinds of compound imagery: integrated and coexistent. In the former mode, the conditioned image made up an integrated form with the perceptual image in one cognitive world. In the latter case, both images (of perception and imagery) coexisted without influencing or interacting with each other (Naruse & Obonai, 1954).

It seems that the clearer the imagery, the more dominant it is in competition; and that the deeper the hypnotic trance in which the conditioning procedure is conducted, the more inhibition of the perceptual image in the observation period.

Now we could understand the above-mentioned facts of interaction between perception and imagery by a schematic representation such as that in Figure 3 which shows the mental mechanism of cognitive monitoring of the perceptual and imaginal images. Figure 3A shows the case of a

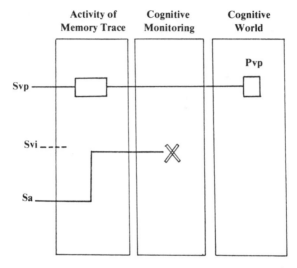

Svp: visual stimulus for perception
Svi: visual stimulus for imagery at the registration
Pvp: visual perception evoked by Svp at the observation
X: blocking loop through cognitive monitoring

Figure 3A. Dominant image of perception inhibits mental image as a result of competition.

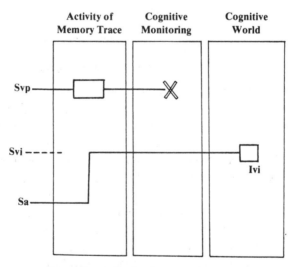

Svp: a visual stimulus for perception
Svi: visual stimulus for imagery at the registration
Sa: auditory stimulus
Ivi: visual imagery evoked by Sa at the observation
X: blocking loop through cognitive monitoring

Figure 3B. Dominant mental image inhibits perceptual image as a result of competition.

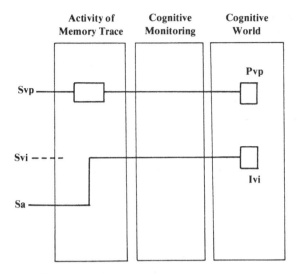

Svp: visual stimulus for perception
Svi: visual stimulus for imagery at the registration
Sa: auditory stimulus
Pvp: visual perception evoked by Svp at the observation
Ivi: visual imagery evoked by Sa at the observation

Figure 3C. Coexisting of mental image and perceptual image without inhibiting each other.

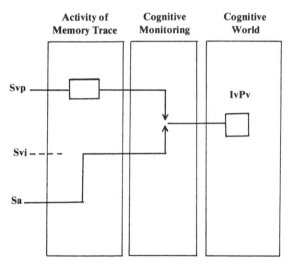

Svp: visual stimulus for perception
Svi: visual stimulus for imagery at the registration
Sa: auditory stimulus
Ivpv: visual image in cognitive world composed with
     Ivi and Pvp

Figure 3D. Integrated form composed with mental image and perceptual image.

dominant perceptual image inhibiting the mental image as a result of competition in cognitive activity. Figure 3B shows the case of a dominant mental image inhibiting the perceptual image through competition between the two images. Figures 3C and 3D represent the mental process of compounding the two images. The former shows coexistence of the mental and perceptual images without inhibition of each other. Figure 3D shows the case in which the subject sees an integrated form or a Gestalt composed of the mental image and the perceptual one.

Leaving the discussion of the mental process of image modification through cognitive monitoring, we will now discuss the interaction of perception and imagery on the subject's awareness or conscious phenomenon. The relationship between perception and imagery, such as that mentioned above, may be represented as shown in Figures 4A, 4B, 4C, and 4D, which indicate the imaginal field and the perceptual field in the subject's cognitive world following a Lewinian scheme. Figure 4A represents the case of a dominant image in the perceptual field which inhibits the mental image in the imaginal field. The solid line means that a recognized image occurs in the cognitive field which inhibits the appearance of an image in the other cognitive field, which is represented by a dotted line. Figure 4B illustrates the case of a dominant mental image in the field of imagery which inhibits the image in the perceptual field. Figure 4C represents coexistent overlapping of the mental and perceptual images in one cognitive world without inhibition or integration. Figure 4D illustrates an integrated form of an image which is composed of both the perceptual and mental images.

The mode of relationship between perception and imagery seems to be determined by the subject's own mental activity in monitoring the cognitive world, by his striving to exclude or include both component images. And the depth of the hypnotic trance preceding image observation seems to be one of the determining factors of the subject's mental activity.

3. Imagery and Meaning

One of the most characteristic modifications in imagery which corresponds to a meaningful but incompletely drawn figural stimulus is the supplementation or completion of an insufficient part of the original figure (Naruse & Obonai, 1953, 1955a; Naruse, 1960a). It is strengthened by verbal suggestion given directly before the observation of imagery. The result of experimental examination of Japanese letter-images indicated that the effect of verbal suggestion or meaning upon image modification was only tentative at the observation period because the subject's mental activity was temporally bound at each trial, and it faded over the passage of time, and then the original shape-effect of stimulus-figure became gradually dominant. Sometimes, the effect of suggestion or meaning disappeared in one successful trial, as if by a kind of tension reduction through observation of the image modified by suggestion or meaning. From the above facts it may be inferred that meaning or a temporal mental set can actively alter the original imagery or make a modified image as a function of cognitive monitoring.

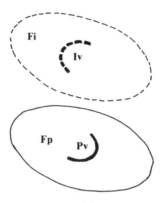

Fi: cognitive field of imagery
Iv: visual imagery inhibited
Fp: cognitive field of perception
Pv: visual perception

Figure 4A. Dominant perception inhibits imagery.

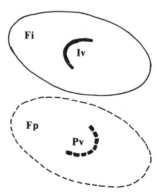

Fi: cognitive field of imagery
Iv: visual imagery
Fp: cognitive field of perception
Pv: visual perception inhibited

Figure 4B. Dominant imagery inhibits perception.

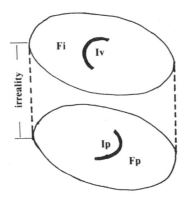

Fi: cognitive field of imagery
Iv: visual imagery
Fp: cognitive field of perception
Ip: visual perception irreality: degree of irreality

Figure 4C. Coexistence of imagery and perception without interaction with each other.

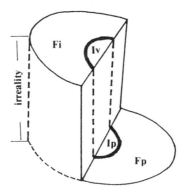

Fi: cognitive field of imagery
Iv: visual imagery
Fp: cognitive field of perception
Ip: visual perception irreality: degree of irreality

Figure 4D. Cognitive integration of perception and imagery.

When such a suggestion with a meaning was given at the original registration period (that is, when a sound was associated with a visual hallucination which was dominated more by hypnotic suggestion than by visual perception) then most subjects at the imagery observation period showed that the meaning-effect by suggestion was more potent in constructing the original image or in registration of a trace of visual experience than the shape-effect produced by a stimulus configuration. An examination of the meaning-effect by the so-called figure-ground reversible figure indicated that the figure-ground relationship of the imagery was generally settled by the subject's original perceptual experience. His anticipatory set imagery observation was able to modify the apparent form of an image. Here also the effect of meaning at registration was more striking in its effect on the appearance and modification of imagery than that of the imagery observation (Naruse & Obonai, 1956).

Mental process as associated with the above-mentioned facts of image modification may be represented as shown in Figures 5A, 5B, 5C and 5D. Figure 5A shows the case of image modification, especially at the stage of cognitive monitoring in which suggested meaning modifies the activity of the original image and helps make up a new form of imagery observed by the subject. Figure 5B shows the case of spontaneous meaning activity at registration. Figure 5C shows the case of a selective determination of imagery by a figure-ground reversible figure which has two incompatible meanings. Figure 5D shows the case of dominant meaning activity by suggestion at the registration period.

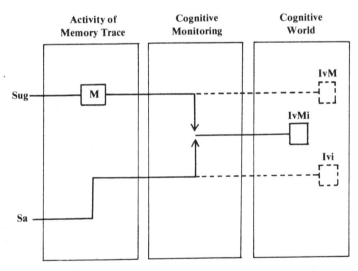

Sug: suggestion verbal
Sa: auditory stimulus
Ivi: visual imagery evoked by Sa at observation
IvM: visual imagery evoked by meaning of verbal suggestion
IvMi: visual imagery composed of Ivm and Ivi
M: meaning activity in organization of memory trace brought
by verbal suggestion

Figure 5A. Suggested meaning activity modifies activity of original image and makes up a new form of imagery observed by subject.

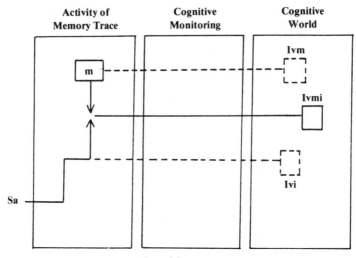

m: spontaneous meaning of figure
Sa: auditory stimulus
Ivi: visual imagery evoked by Sa at the observation
Ivmi: visual imagery composed of Ivm and Ivi
Ivm: visual imagery evoked by spontaneous meaning taken hold at
      the registration period

Figure 5B. Spontaneous meaning activity of Pv (original visual perception) modifies activity of Tv (memory trace of Pv) at the registration period.

4. Posthypnotic Facilitation of Imagery

The above mentioned type of experimentation has been performed not only by the author but also by Leuba (1940, 1941, 1955, 1957), Leuba and Dunlap (1951) and Fisher (1955). All of these investigators reported results that were in basic agreement. They, nevertheless, failed to obtain agreement on their interpretation of the nature of this imagery.

Leuba regarded the imagery as a conditioned sensation which was facilitated during hypnosis because of a relatively complete concentration on the conditioned and unconditioned stimuli and the consequent absence of conflicting and inhibitory factors at the time of conditioning. Stating that these responses did show a marked similarity to behavior induced by explicit posthypnotic suggestions, Fisher criticized Leuba's theory as the "alleged conditioning phenomena," and emphasized a conative factor or role-taking as an essential condition of the phenomenon. In spite of such disagreement, there seems to be a congruity between both investigators that the evoked experience is a sort of perception.

According to Naruse (1962a, 1965a), however, it would be better to classify this phenomenon not as a real sensation or perception but as a kind of imagery or hallucination. It is initiated through a conditioning

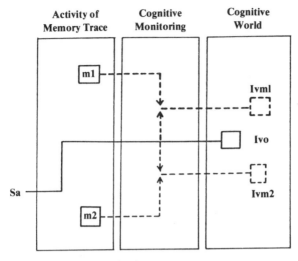

Activity of   Cognitive   Cognitive
Memory Trace   Monitoring   World

Sa: auditory stimulus
m1: one side meaning of f-g reversible figure
m2: another side meaning of f-g reversible figure
Ivo: visual imagery of original figure by Sa
Ivm1: visual imagery of Ivo modified by dominant
      activity of m1 through cognitive monitoring
Ivm2: visual imagery of Ivo modified by activity of
      m2 through cognitive monitoring

Figure 5C. Selective determination of imagery by figure-ground reversal figure which has two incompatible meanings.

procedure, but the evoked response is too faint or too weak to appear as an image in the ordinary waking state. To be recognized as an image, it would have to be strengthened by a facilitating condition, such as that provided by the hypnotic trance state.

Now we shall examine the nature of imagery in this type of experiment. The basic procedure is shown in Figure 6A. The conditioning procedure in the registration period was conducted in the hypnotic trance in which the subject was brought into as deep a trance as possible. Then he was given a posthypnotic suggestion, that he would forget completely the whole experience at the registration so that he could not remember it even after being awakened from the trance. After confirming the amnesia procedure of the registration, he was awakened and it was confirmed again that posthypnotic amnesia of the registration occurred in the waking state. Thereafter, the procedure of imagery observation by conditioning stimuli was given again. When such a procedure is applied without hypnosis, the subject's experience may be generally called a memory image because the subject is able to recognize it as a familiar experience which has occurred previously. On the contrary, in the above-mentioned experiment, the

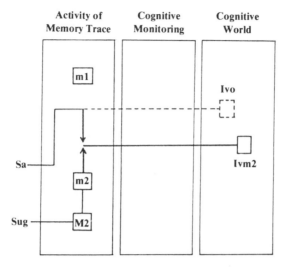

Sa: auditory stimulus
ml: one side meaning of f-g reversible figure
m2: another side meaning of f-g reversible figure
Sug: suggestion verbal
M2: meaning activity of suggestion which strengthens
      activity of m2 at the registration
Ivo: visual imagery of original figure by Sa
Ivm2: modified imagery of Ivo by m2 strengthened by M2

Figure 5D. Dominant imagery evoked by one side meaning of figure-ground reversible figure strengthened by meaning of suggestion at the registration.

subject is not able to have an awareness of such an identification (recognition). That is, when an image from a past perception (or memory image) comes to mind with presentation of the conditioning stimulus, the subject cognizes it as a new experience or phenomenon like a perceptual image. Even though some subjects believed it was real perceptual image, some wondered whether it was a visual image or if it was only an idea or thought. And some others were certain they saw a visual image, but they believed that there was no real stimulus corresponding to such an image.

In everyday life it is very rare to encounter a situation in which it is difficult to differentiate an image as perceptual or imaginal, and also in which the subject is unsure whether or not there is an objective stimulus corresponding to his conscious imagery. Even though such a situation is so rare and abnormal in everyday life, it is a well-known phenomenon in the hypnotic trance. Then, Naruse and Obonai (1953-1956) inferred that at the period of imagery observation the subject might not be in the normal waking state but in a kind of abnormal or trance state such that conditioned imagery could appear. In fact, some subjects in the period of imagery observation displayed some very specific behaviors which were known as indications of the trance state; these were closing the eyes, turning pale, an inactive or lethargic face and action, a vacant face, indifferent behavior to reality, etc. The authors considered that the conditioning stimulus was a cue not only for the original sensory image but also for the primary hypnotic state itself in which the conditioning had been performed. Such a conditioned or secondary trance the authors characterized as "posthypnotic hallucinatory state." In those days the authors assumed that such a trance-like state might be an essential and necessary condition for the experience of imagery. Figure 6B shows the mental process in which a secondary trance was spontaneously induced by a conditioned cue in the observation period of imagery.

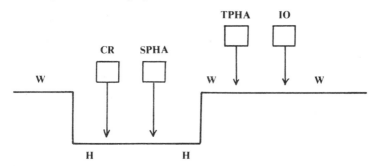

Figure 6A. Experimental procedure of conditioned imagery.

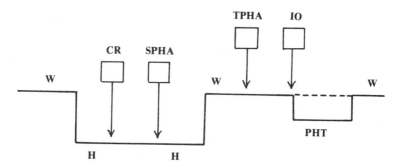

W: waking state
H: hypnotic state
CR: conditioning procedure at the registration
SPHA: suggestion of posthypnotic amnesia in hypnotic state
TPHA: testing of posthypnotic amnesia in waking state
IO: testing procedure of imagery at the observation
PHT: posthypnotic trance induced by Sa (auditory stimulus) at the observation of imagery

Figure 6B. Posthypnotic trance induced at imagery observation.

As research progressed, however, the author (1962a, 1965a, 1969) became aware that it was not necessary to assume that a special mental state such as the hypnotic trance was an essential condition for the appearance of conditioned imagery. This was because most subjects reported imagery in the observation period without any indication of hypnotic behavior. And also, subjects who did see imagery with hypnotic indications in the beginning of the experiment, in the progress of experimentation, gradually came to report imagery without any trance indications during the observations. However, it seems certain that if the conditioned trance does occur at the observation, it facilitates or strengthens imaginal mental activity.

II. IMAGERY IN HYPNOTIC TRANCE

In defining the depth of trance by the appearance of positive or negative hallucinations, we have recognized hallucination as one of the most common or typical determinant indices of hypnosis. In this context, it is not possible for us to refuse Orne's (1962) proposition that "all hypnotic phenomena are either hallucination or intimately related to them."

In regard to the concept of hallucination, we have to consider differentiating two contexts: (1) imagery is vivid, and (2) imagery is perception-like in experience. As to imagery in hypnosis, two kinds are differentiated from the viewpoint of the subject's activity: suggested and spontaneous. The former is most usually seen in the process of trance induction and is used for the purpose of controlling the experimental condition by verbal suggestion. In responding to a suggestion by the hypnotist or experimenter, some image comes to the subject's mind. And the image represents directly, or with some modification, the meaning of the suggestion. In such a case the subject usually stays passive, inactive or absent-minded following the active suggestion by the hypnotist or experimenter.

On the contrary, spontaneous imagery is seen typically in situations involving so-called hypnotherapy, especially in hypnoanalysis. In this type of situation, mental imagery appears to the patient spontaneously through free association or emotional or affective excitement during the process of psychotherapy. Necessary conditions for the appearance of spontaneous imagery are positive mental activity by the patient or subject and positive helping and encouragement of the patient without active suggestion by the therapist, experimenter or hypnotist. In such conditions, the subject is able to make up a mental image more freely or creatively according to a particular idea, thought, emotion or trend in mind without restriction or inhibition from the self or some other person.

Another aspect of hypnotic hallucinations is that it is not easy for hypnotic subjects to discriminate their own imagery from real perception, that is, subjects experience an image as an apparent perception of an external object when no such object is present. Even if such experience is liable to be associated with pathological conditions such as acute organic delirium or schizophrenia, we may point out that there is also a high potential for confusion in normal individuals under certain conditions. Hypnosis is one condition in which the probability of such a confusion is very high.

In the everyday-life waking state, a person usually tries to attend restrictively to perceptual images corresponding to objects in the outer world without any confusion like the above-mentioned. However, when imagery becomes so vivid and definite in the hypnotic trance, perceptual images and imaginal images are incompatible with each other, or integrated into a well-constructed form in vision which is composed of both imagery and perception. Such phenomena may also be encountered in the eidetic imagery of waking consciousness (Ahsen, 1977; Marks & McKellar, 1982). Sometimes, both types of imagery (mental and perceptual) coexist independently of each other. Here we may infer that one of the necessary conditions for a hallucinatory experience surely is to make the subject switch over or convert his mental set from the perceptual level to a hallucinatory one by blocking out a sensory input or inhibiting an active attention to reality or the objective world.

In addition to a conversion of mental set, furthermore, when the subject is absorbed in inner activity, an hallucination appears and develops spontaneously, freely, and creatively. Such absorption seems to be characteristic of the hypnotic trance, especially in hypnotherapy.

The ability to be induced into a hypnotic trance which seems to be a kind of facilitator or accelerator of hallucinatory experience is termed "hynotizability" or "suggestibility." There is a large literature on individual differences in hypnotizability. It is certainly true that some subjects are able to have very vivid hallucinations on the first trial, while others are not able to even after several trials. However, it is also a well-known fact that by repetition of suitable procedures of induction and suggestion under optimal conditions, subjects become able to have positive experiences of imagery or hallucinations and become better hypnotic subjects.

Furthermore, through a process of repetitive induction or experience of the hypnotic trance, the subject usually changes his pattern of behavior in trance. At the stage of being a beginner as a hypnotic subject, the subject shows passive, inactive, sleepy, trance-like behavior following the hypnotist's action and suggestion. Some beginners behave at times in a dramatic pattern typical of hypnotic phenomena. However, when more familiar with hypnotic experiences after repeated sessions, the subject behaves as in everyday life, and as usual in the waking state. Sometimes the subject shows dramatic behavior, but at other times behaves by his own will, responding to the same hypnotic suggestion by the hypnotist. According to his introspection, the subject is aware of everything happening in the situation, and, with a moderately awake and rational mental set, selectively decides the most suitable behavior. If the subject chooses to follow the suggestions, he is able to go into the suggested world and behaves according to his imaginal life. And when the subject makes the decision to carry out his own intentions, he is able to do so, disregarding the suggestions of the hypnotist. Now, the subject is free to choose or decide to behave or experience at the imaginal level or the perceptual level of cognition, and/or the unreality or reality level of the living world. This means that such an ability is acquired through the repeated experience of cognitive activity along the perception-imagery continuum and mental states along the waking-trance continuum.

III. HETERO-HYPNOSIS AND SELF-TRANCE

1. Autogenic Training as Self-Trance

Having applied autogenic training as constructed by J.H. Schultz (1932) to patients and trainees since 1955, the author has found that for Japanese people it is more useful than psychodynamic methods as a psychotherapeutic procedure or training for everyday living. As appears to be true of other countries in the Orient, self-therapeutic methods seem more useful for the Japanese than the European hetero-therapeutic ones. In differentiating hypnosis into hetero-hypnosis and self-hypnosis, the author (1962) assumed that autogenic training is a technique which brings the trainee into a trance of self-hypnosis which is induced through the self-striving of the trainee. The goal of this striving is to realize a series of formulae which are composed of waking suggestions. The realization involves the task of obtaining hallucinatory experiences of imagery indicated by a formula, which is a kind of self-suggestion.

For such realization, the subject has to learn the knack of acquiring the right mental set through repetitive, everyday exercise. As the most important point, Schultz and Luthe (1959) stated that passive concentration is necessary. And Naruse (1962a) characterized such a mental set as a state of meditative concentration in which the passive concentration was so well generalized that a unique and hallucinatory mental set was firmly made up without any disturbance from external stimuli or reality.

Now let's turn our attention to the training process. Summoning up hallucinatory experiences according to a formula or realization of self-suggestion requires successful exercise of the training. For example, if a trainee gets an experience of warmth in his arm through striving to concentrate on a formula, "My arm is warm," then the exercise is progressing well. Harano, Ogawa and Naruse (1965) proposed that there were two important conditions for the realization of a formula in this training: *attention* and *striving*. The former means the so-called mental contact with the part of body indicated by the formula and also maintenance of awareness of the bodily part. We may characterize striving as the subject's effort to realize the formula in his mind, that is, to get an experience of warmth in his arm(s).

Some subjects who are beginners in such training are liable to concentrate so actively to realize the formula that the experience of warmth is difficult to obtain. Even though the suggestion is relized in their subjective experience, physiological responses still appear in a contrary way to the self-suggestion. That is, it is found that active striving gives rise to a diminution of finger pulse volume, pulsewaves, fluctuations of the base-line of the pulsewaves, and a decrease in skin temperature at the finger. From this fact, it may be inferred that active striving to the formula (self-suggestion) of warmth provokes a heightening of the excitatory level of the autonomic nervous system.

On the contrary, a subject who has mastered the training well is now able to strive in both ways at will, that is, actively and passively, according to the request of the trainer or experimenter. Through passive striving, the effect of suggestion appears subjectively and also objectively (physiologically) in the most skilled trainee; that is, an increase occurs in the amplitude of the volume pulse with a rise of skin termperature. This fact verifies that vasodilation, or an increase in blood volume, has occurred and that the suggestion has successfully increased skin temperature. When a skilled trainee tries active striving for warmth, failure to obtain a positive effect is indicated not only subjectively but also objectively, that is, a fall of skin temperature is accompanied by a monophasic vasoconstriction, that is, a temporary decrease in blood volume.

In spite of the fact that autogenic training does not claim physiological realization of a formula but realization only in subjective experience, the above-mentioned responses of the human organism as a whole including subjective and objective realization of the suggestions seem to provide a good indicator of the mechanism of the subject's striving.

2. Hetero-Hypnosis and Self-Hypnosis

In addition to the passive concentration described by Luthe and Schultz,

Naruse (1962) stressed a specific mental set for a successful exercise in which the subject strives to be indifferent to or dissociate from the realistic, rationalistic, logical and objective world. Naruse states that an essential characteristic of meditative concentration is the subject's striving to devote himself to, or be absorbed in, imaginal activity with a positive mental set to accept and go into unreality, irrationality, or the illogical and subjective world. Such a mental state as the above-mentioned, generally called a trance, the author has named a state of "meditative concentration."

Even if the trance brought about by autogenic training or self-hypnosis seems not to be different in its essential nature then the trance induced by covnventional hypnosis, that is, hetero-hypnosis, in the author's view there are some important differences between the two. One of them is the degree of absorption, and the other is the interrelationship with the hypnotist or trainer.

In the case of hetero-hypnosis, the subject is able to depend completely upon the hypnotist without any concern for outside circumstances, and the hypnotist has to try to take every care and responsibility in the hypnotic situation. Therefore, the subject is permitted to try to follow only the hypnotist and to realize positively all of the hetero-suggestions without any consideration for reality and the objective world. On the contrary, in the case of self-trance, the subject has to take two different kinds of role, that is, as both hypnotic subject and hypnotist, or trainee and trainer, because he has no other person to safeguard the trance. He has to be alert and take the responsibility for everything during self-trance. Since he has no time to be absorbed in imaginal activity completely, as in the case of hetero-hypnosis, then he has to control and hold himself at each moment to be adaptive to objective circumstances even in the trance. This is the reason that self-trance, like autogenic training, is generally not as deep a trance as hetero-hypnosis. Then we may conclude that even if the imagery experience is very active, the hallucinatory image is not so easy to form as it is in hetero-hypnosis, which makes the hallucinatory experience so dominant.

3. The Neutral Trance and the Disturbed Trance

Differentiating two types of trance - (1) neutral hypnosis or pure trance and (2) agitated hypnosis or disturbed trance - Naruse (1962a) pointed out that the former type is more possible to induce through a self-hypnotic procedure such as autogenic training, and the latter type is usually induced by the hetero-hypnotic procedure.

In the induction technique of hetero-hypnosis, it is necessary for the hypnotist to utilize every possible condition to disturb or agitate the subject's mind or affect for the purpose of the induction, deepening the hypnosis, and strengthening the effect of suggestions of hypnotic phenomena. Concerning these conditions, many authors in the hypnosis literature have described this process as goal-directed striving, role playing, demand characteristics, task motivation, transference, or in terms of the interpersonal relationship, psychodynamic or psychoanalytic conceptions and so on. The more disturbed or agitated the trance, the more the subject is absorbed in the imaginal world.

On the contrary, in the process of self-hypnosis, the trainee has to strive solely to concentrate passively and go into a state of meditative concentration with self-control to achieve less absorption or disturbance and agitation than occurs in hetero-hypnosis. His sole concern is to try to relax, to get sleepy, to enter an altered state of awareness, with a feeling of discontinuity in consciousness, functional dissociation, an indifferent mental set, an autistic attitude, readiness for imagination, acceptance of unreality and so on, without directive influence from another person.

The author has already mentioned the case in which a subject changed from a passive and inactive behavior at the first trial of induction to an active behavior according with his own intentions, free from hetero-suggestion. it is a commonly perceived fact that the experience of a subject who becomes familiar with the hypnotic experience through repeated induction seems to become gradually refined, from a disturbed and agitated hypnosis to a neutral hynosis. However, it seems to be true only when the hypnotist is not so active as to disturb his subject, treats his subject in a client-centered fashion and encourages the subject's own activity in the trance. Such a case of hetero-hypnosis seems to be not different from that of self-hypnosis from the viewpoint of neutral hypnosis or pure trance in which the subject is free from suggestions from, or a human relationship with, the hypnotist and is able to follow his own mental processes.

Even though the subject is able to behave by his intentions, he may still be disturbed by his own inner instability or affective disturbance in a manner similar to neurotic or psychotic conditions. Therefore, a pure trance may be brought to him only when he is stable in emotion and calm in mind. The goal of zen meditation or yoga training may be this kind of trance. As a practical method for inducing a pure trance, the author proposes the systematic method of "Jiko-control" in the latter part of the present chapter.

IV. IMAGERY IN WAKING

1. Imagery in the Waking State

It is a generally perceived fact that the experience of imagery or imaginal mental activity is not very distinctive or vivid in the waking state. Sometimes hallucinatory experience has been reported spontaneously in everyday waking life. However, critical study indicates that it is usually found not in "waking" mental states but in trance-like states such as drowsiness, sleepiness or daydreaming, or in the hypnagogic or hypnopomic state. But, it is not impossible to have imagery in the waking state. Mental imagery in the waking state does not usually appear spontaneously, but is called into awareness or is made up intentionally by striving to do so for visual cognition, problem-solving, thinking, recalling, recognition, fantasy, artistic or creative activity and so on.

In psychotherapeutic sessions or in the process of free association, word association, or word-imagery association, the psychotherapist is familiar with the spontaneous appearance of imagery which is sometimes vague faint and apt to disappear and sometimes vivid and clear like a photograph or a motion picture film. In the former case, the patient is not sure whether he sees imagery certainly or only experiences an idea without imagery. In the latter case, he is involved in some imaginal situation or emotional story through imaginal mental activity.

Relying on these experiences of imagery in the waking state, the author and his colleagues tried to get a patient to see some appointed imagery intentionally during psychotherapy. Sometimes a light trance with hypnosis was utilized; sometimes the patient was asked to strive as in autogenic training; at other times imagery was seen directly in the waking state without any utilization of the trance-like state. Successful results from each of these conditions indicated that even in the waking state a patient or a subject may not find it so difficult to experience intentional imagery, especially in visual cognition, through a specially planned exercise for imaginal mental activity.

2. Systematic Exercise for Imagery

Laying stress on the therapeutic effect of self-striving of the patient by himself, the author (1982) and his colleagues have tried to develop a systematic method of "Jiko-control" which grew out of the basis of autogenic training and Jacobson's progressive relaxation, and the above-mentioned training of a waking subject for intended imagery. Although the Japanese word "Jiko" is translated as "Self" in English, Jiko-control should not be confused with the self-control concept associated with behavior therapy. Jiko-control instead represents the view that all psychotherapy is essentially self-therapy, thereby differing fundamentally from the hetero-therapeutic point of view such as psychoanalysis and most other Western psychotherapies. And also it has been discovered that imaginal mental activity becomes more reliable for intended imagery through the training of Jiko-control in a waking subject, and that the training is very useful for clinical and experimental applications of imagery.

The training system of Jiko-control is composed of five stages of self-exercise: (1) relaxation; (2) meditation; (3) imagery; (4) self-understanding; and (5) control of action.

V. IMAGINAL MENTAL ACTIVITY

1. Task Striving

Now we shall present a theoretical discussion of the genesis and modification of imagery in general.

In the conditioned imagery experiment there were some subjects who did not have any hallucinatory experience when a sound was given at the first trial after the conditioning procedure in hypnosis. Some of them began successfully to see imagery by drawing the image when the following instruction was given just before the observation period: "You may see something on the card"; "You will be aware of something visual"; or "Try to concentrate to something visual in your blank mind" and so on. The above fact seems to indicate that the subject's own striving to see or get a hallucination or image, or his mental set for the imaginal experience, is a necessary condition for the appearance of conditioned imagery.

For each and every phenomenon and behavior specific to hypnosis, the motivation or conative factor had been stressed as essential to the nature of hypnosis by several authors; for instance, goal-directed striving by White (1941), role playing by Sarbin (1950), task motivation by Barber (1969), and demand characteristics by Orne (1969).

Concerning the phenomena of hypnotic hallucination and imagery, the author is convinced that the subject must use his own striving to get them into his mind, consciously or subconsciously. In addition to intentional imagery, it has been discovered that spontaneous hallucinations and even suggested ones were made up or modified according to the subject's voluntary activity, and also that his mental set selects or decides the imaginal level in between imagery and hallucination. Making up what he does see in imagery and the decision as to how he does see the imagery may be seen as a task for the subject. If so, we may now infer that the appearance and modification of hallucinations or images in hypnosis are the result of task striving by the subject which is characterized by a state of meditative concentration.

We noticed also how the subject's own striving is an essential condition for getting a hallucinatory experience or intended imagery by autogenic training or self-hypnosis in which passive or meditative concentration is necessary for the trainee. As stated in the present paper, imaginal mental activity becomes more positive and free for the subject in the waking state through repetitive experience or a purposeful exercise to get imagery.

In imaginal mental activity, it may be reasonable to discuss the task which the subject strives to perform by differentiating it into two categories: image construction and imaginal mental set.

2. Image Construction

One of the tasks which the subject has to strive to complete in his imaginal mental activity is the construction or manufacture of imagery in his cognitive field. The material for image construction is his past experience stored in his memorial organization on the one hand and his present situation, both inner and outer worlds, on the other.

As the author stated in regard to conditioned imagery, mental images appear in the subject's cognitive world through his striving to bring them there. Since the origin of the mental image is thought to be an activation of memory traces of past experience, then we may call it an *original image,* which is only an hypothetical construct and does not appear directly in awareness. When the original image is brought into awareness as conscious imagery, it is sometimes modified and sometimes not. And the subject's striving seems to decide the mode of modification. Composition, decomposition, modification by meaning, and competition between perception and imagery were mentioned as instances of image construction earlier in the present paper. Imagery produced by suggestion and intention was mentioned as the result of the subject's striving to create it according to the indication of its meaning.

The subject's attempt to construct an image in conscious awareness using the original image which is also constructed from the materials of inner and outer activities may be characterized as *cognitive monitoring* in imaginal mental activity. Usually cognitive monitoring seems to work voluntarily and intentionally but out of awareness, and the subject experiences the imagery as it comes, spontaneously or as natural result, but he is not aware that he made it up by himself, constructing it with his own inner and outer materials and his own intention.

3. Imaginal Mental Set

The other purpose of the tasks involving the subject's striving in imaginal mental activity is to bring the constructed imagery into awareness or consciousness, or to make a preparatory condition which facilitates the apperance of imagery in cognitive awareness or consciousness. Striving to facilitate awareness of imagery may be characterized by an imaginal mental set which finds a way of confronting the subjective, inner or unreal world on one side and the objective, outer or real world on the other side. Sometimes, it indicates the mode of attention or concentration and sometimes the degree of alteration in the state of consciousness.

An effective imaginal mental set is indicated by the following: mental relaxation, passive concentration, dissociation from the perceptual world, meditative concentration, acceptance of imaginal experience, positive adjustment to the imaginal world, absorption into imaginal experience, staying in a state of meditative or altered consciousness and so on.

The hypnotic trance or self-induced, trance-like state is seen as typical of the stabilized mental set for imagery in which imagery is apt to appear as an hallucinatory experience, or comes to the subject as a very clear and vivid mental image. In autogenic training or by an exercise to get imagery, the imaginal mental set becomes more firm and positive.

4. Extension of Cognitive and Mental Activities

Imaginal activity is very general and important in human mental activity. But it is not so active and powerful that it cannot be inhibited or disturbed by perceptual activity which is dominant in cognitive activity. If it would become as powerful and definite as perceptual activity and if it could work equally with perceptual activity, then cognitive activity could be enriched and become more free.

In the hypnotic trance imaginal activity becomes too dominant, and is apt to inhibit perceptual activity and then to disturb well-balanced cognition. But, as mentioned already, when the subject becomes familiar with hypnotic behavior through repeated experience, then he is able to choose freely between imaginal and perceptual mental activities according to his own preference, that is, sometimes perceptual only, sometime imaginal only (hallucinatory), and at other times both activities cooperatively, and so on.

By exercises of imaginal activity such as autogenic training or Jiko-control, the trainee becomes capable of extending his cognitive activity in the above-mentioned manner. The goal of Japanese zen meditation or Indian Yoga training is said to be the same as that of Jiko-control: that is, not only the extension of cognitive activity but also the promotion of mental activity in general.

REFERENCES

Ahsen, A. (1977). Eidetics: An overview. *Journal of Mental Imagery, 1*(1), 5-38.
Barber, T.X. (1969). *Hypnosis: A scientific approach.* New York: Van Nostrand.

Fisher, S. An investigation of alleged conditioning phenomena under hypnosis. *Journal of Clinical and Experimental Hypnosis 3*, 71-103.

Harano, H. Ogawa, K., & Naruse, G. (1965). A study of plethysmography and skin temperature during active concentration and autogenic exercise. In W. Luthe (Ed.) *Autogenes Training.* Stuttgart: George Thieme, 55-58.

Jacobson, E. (1929). *Progressive relaxation.* Chicago: University Chicago Press.

Leuba, C. (1940). Images as conditioned sensations. *Journal of Experimental Psychology, 26*, 345-357.

Leuba, C. (1941). The use of hypnosis for controlling variables in psychological experiments. *Journal of Abnormal and Social Psychology, 36*, 271-274.

Leuba, C. (1955). Conditioning during hypnosis. *Journal of Clinical Hypnosis, 33*, 256-259.

Leuba, C. (1957). The reality of hypnotic phenomena: A critique of the role playing theory of hypnosis. *Journal of Clinical and Experimental Hypnosis, 5*, 32-38.

Leuba, C., & Dunlap, R. (1951). Conditioning imagery. *Journal of Experimental Psychology, 41*, 345-351.

Marks, D., & McKellar, P. (1982). The nature and function of eidetic imagery. *Journal of Mental Imagery, 6*(1), 1-124.

Naruse, G. (1960a). The abstract image in the post hypnotic state. *International Journal of Clinical and Experimental Hypnosis, 8*, 213-229.

Naruse, G. (1960b). *Hypnosis* (in Japanese). Tokyo: Seishin.

Naruse, G. (1962a). Hypnosis as a meditative concentration and its relationship to the perceptual process. In M.V. Kline (Ed.) *The nature of hypnosis.* Baltimore: Waverly Press, Pp. 37-55.

Naruse, G. (1962b). A contribution to systematic understanding of modification in post-hypnotic hallucination. *Japanese Journal of Educational and Social Psychology, 3*, 40-53.

Naruse, G. (1965a). A genetic consideration of hypnotic hallucination. *Japanese Journal of Educational and Social Psychology, 5*, 15-19.

Naruse, G. (1965b). The hypnotic treatment of stage-fright in champion athletes. *International Journal of Clinical and Experimental Hypnosis, 8*, 63-70.

Naruse, G. (1969). *Jiko-control* (in Japanese). Tokyo: Kodansha.

Naruse, G. (1974). On the development of Educational Hypnosis. *Research Bulletin of Educational Psychology, Faculty of Education, Kyushu University, 18*,(1) 11-16.

Naruse, G. (1975). On the application of hypnosis to sport. In L. Unestahl (Ed.), *Hypnosis in the seventies.* Orebro: Welins Tryckeri Eftr. Pp. 171-175.

Naruse, G. (1982). A systemic method of Jiko-control. In J.K. Zeig (Ed.) *Ericksonian approaches to hypnosis and psychotherapy.* New York: Brunner Mazel, Pp. 391-398

Naruse, G. (1983). On the training of imaginal mental activity. *International Imagery Bulletin.*

Naruse, G., & Obonai, T. (1953). Decomposition and fusion of mental image in the drowsy and post-hypnotic hallucinatory state. *Journal of Clinical and Experimental Hypnosis, 1*(3) 23-41.

Naruse, G., & Obonai, T. (1954). Decomposition and fusion of mental images in the post-hypnotic hallucinatory state. III. The influence of perception of mental image. *Japanese Psychological Research, 1*, 21-33.

Naruse, G., & Obonai, T. (1955a). Decomposition and fusion of mental images in the post-hypnotic hallucinatory state. II. Mechanism of image composing activity. *Journal of Clinical and Experimental Hypnosis, 3*, 2-23.

Naruse, G., & Obonai, T. (1955b). Decomposition and fusion of mental images in the post-hypnotic hallucinatory state. IV. On the interaction of mental images in succession (in Japanese). *Japanese Journal of Psychology, 25*, 246-255.

Naruse, G., & Obonai, T. (1956). Figure-Ground image reversal in the post-hypnotic hallucinatory state. *Japanese Psychological Research, 4*, 1-10.

Naruse, G., & Schultz, J.H. (1963). *Self-hypnosis* (in Japanese), Tokyo: Seishin.

Orne, M.T. (1962). Hypnotically induced hallucination. In L.J. West (Ed.) *Hallucination,* New York: Grune & Stratton.

Orne, M.T. (1969). Demand characteristics and the concept of quasi-controls. In R. Rosenthal and R.L. Rownow (Eds.) *Artifact in behavioral research.* New York: Academic Press. Pp. 143-179.

Sarbin, T.R. (1950). Contribution to role-taking theory. I. Hypnotic behavior. *Psychological Review, 57,* 255-270.

Schultz, J.H. (1932). *Das Autogene Training.* Stuttgart: George Thieme.

Schultz, J.H., & Luthe, W. (1959). *Autogenic training.* New York: Grune & Stratton.

White, R.W. (1941). A preface to the theory of hypnotism. *Journal of Abnormal and Social Psychology, 36,* 279-289.

Learning and Imagery

B. R. Bugelski

There is no point to dividing learning into classical (Pavlovian) and instrumental (Thorndikian, Skinnerian) types. All associations are necessarily formed between neural events. Images are conditioned responses (sensory neural activities) which normally occur on the occasion of some CS. One cannot decide to have an image-whatever one "decides" to image will already be determined by prior events (external or internal stimulation). Images occur in any conditioning situation prior to the occurrence of any overt response. Thus, all imagery is conditioning and all conditioning is the formation of images with or without "reportable" accompanying subjective experience. Proprioceptive stimulation occurring as a CS can serve to initiate action which it formerly followed. Such

This chapter was originally prepared for Open Peer Commentary in the *Journal of Mental Imagery*, 1982, *6*, 1-92. Seventeen psychologists reacted to the paper with varying degrees of approval and disapproval. The present chapter is a revision of the original presentation in the light of the critical commentaries. The author is indebted to Dr. Erwin Segal who read a preliminary version of this paper. Dr. Segal suggested the treatment of epiphenomena, which appears late in the paper, as correlates of causal chains.

kinesthetic activity is qualitatively the same as imagery although rarely described in subjective terms. The conditions governing the formation of images are the same as the conditions for learning: (1) the contiguous occurrence of two neural events over a sufficient period of time, and (2) the lack of any other (interfering) stimulation. Subjective experiences of imagery, sometimes derogated as "epiphenomena," can serve useful functions as correlates of underlying neural events in causal chains. Psychotherapists can perhaps benefit from controlling (reconditioning) patients' imagery if they can identify the significant images aroused by their basic tools, namely, words.

Prefatory Note: Before I can hope to communicate with the generation of psychologists who came into the field in the sixties and seventies and their students, I must offer this preliminary note. This paper amounts to what might be considered an elementary lecture on the subject of imagery. It will probably be regarded as old-fashioned, perhaps even passe, as it will not employ the language of modern cognitive psychology which the author regards as prematurely rejecting the behavioral approach that showed such promise in the period just before World War II. There will be no talk about propositions, representations, mental structures (deep or shallow), compartmentalized and suitably boxed memories, coding, schemas, searches (Who does the searching, by the way?). The great psychologists of the thirties and forties have not received the respect due them in modern presentations of psychological problems. Graduate students no longer hear of Hull and Tolman, to say nothing of Guthrie. The great Pavlov has been relegated to some historical niche that is populated by salivating dogs, and a whole generation of young psychologists is going about studying the mind with computers but is never exposed to the mind-body problem. Nativism has invaded the field anew, with innate grammars and computational processes performed by homunculi with mental eyes. Behaviorism is regarded as a dead horse, but I believe the horse of behaviorism has not died. He has been suffering an exile, and this paper is an attempt to return him to his proper place at the head of the psychological parade. There are no new or original ideas in this paper. Everything in it can be regarded as an exploitation of the contributions of Clark Hull, O. H. Mowrer, Donald Hebb, and A. C. Greenwald. They, of course, are not responsible for any of my interpretations of their thinking.

For a little more than a decade now we have seen the revival and thriving growth of interest in imagery. A great deal of research in areas ranging from perceptual effects through mnemonics and memory to behavior therapy and education has been published, and to some extent, enjoyed. But the new interest in imagery had hardly begun to blossom when controversy arose. Anderson and Bower (1973) questioned the "picture metaphor" and opted for a different approach to cognitive problems via "propositions," Pylyshyn (1973) added to the attack, not so much challenging the reality of imagery as assigning it the ephemeral status of an epiphenomenon. Kosslyn and Pomerantz (1977) came to the defense of imagery as a functional variable of great potential and Anderson (1978) returned to the lists with the suggestion that the controversy was unresolvable.

It is quite likely that the controversy as expressed in the debates mentoned is unresolvable where images are defined by each side to suit the debater's arguments. It is the purpose of this paper to present a somewhat different analysis of the nature of imagery which would make the controversies superfluous.

At the outset I should point out that Donald King (1979) and Akhter Ahsen (1979, 1982) have already provided a theoretical analysis very similar to the one I shall present. They have made imagery a function of learning. I shall do likewise but, at the same time, I shall try to suggest that learning is a corresponding function of imagery. They have extended their theories to a variety of perceptual and memory functions and operations. My immediate purpose is to address the question of the basic nature of imagery itself before venturing into extensions. With respect to the issue of instrumental learning and voluntary behavior, I believe that King has not clarified sufficiently the central role of imagery, and that will be the major emphasis of this paper.

I. THE STATUS OF THE LEARNING PROBLEM

In the pre-World War II period, which most of our present-day psychologists may know feebly from casual exposure to history, the key questions in psychology were: What is learning? How does it occur? What is its role in behavior? Intellectual battles were fought over the importance or significance of drive, purpose, reward, expectancy, and contiguity. Psychologists took sides: Some lined up with Hull, others with Tolman, still others with Skinner, old hands stuck with Pavlov and Thorndike. The battles were waged in classrooms, journals, books, in bars, and at meetings. Some armies confronted each other, some ignored the field and protected their own turf, refusing to recognize the enemy. With the termination of World War II, changes began to occur as the old generals died off; clinical psychology was born and multiplied in the number of various splits and divisions of therapeutic practice. Learning, as an area, began to lose in interest as the majority of graduate students concentrated on therapy. Today, courses in learning are becoming rare. The behavior of rats came to appear as a less than satisfactory model for psychologists. The failure of nonsense syllables to provide analytical insights spurred some learning psychologists to experiment more boldly with words. Free recall was born and the personal contributions of subjects to the learning process began to take center stage. Cognitive psychology began to replace the old behaviorism. The influence of linguists began to be felt in the new linguistics and psycholinguistics. The "mind" became respectable again and computer models of memory came to be the driving concern. Never mind that before one can remember he must learn. The learning process was ignored, as much of what was learned by an adult human working with words seemed to be acquired in one trial. The dull repetitions of Ebbinghausian nonsense syllables gave way to single presentations of word lists, phrases, sentences, and connected discourse. "How much is retained?" became the question. How one gradually arrived at perfect recall lost favor as a question.

1. The Preoccupation with Behavior and Overt Responses

Why did learning lose its status as a problem? Why are Hull, Tolman, and Guthrie no longer respected and esteemed? A glance at the titles of their books (a glance at the titles will do as no one reads the books anymore) suggests an answer. They, too, were not really studying learning. Except for Guthrie, whose major contribution was called the *The Psychology of Learning* (1935), the other titans called their books: *The Principles of Behavior* (Hull, 1943), *The Essentials of Behavior* (Hull, 1951), *The Behavior of Organisms* (Skinner, 1938), *Purposive Behavior* (Tolman, 1932), etc. Even Hebb (1949) entitled his masterwork *The Organization of Behavior*. These psychologists were concerned with behavior and not learning even though they were commonly called learning psychologists. In his original *Theories of Learning*, Hilgard (1956) defined learning in terms of what it is not and suggested that students wanting to know what learning is should look at what the workers in the field were doing. Learning became what learning psychologists study, a nice operational definition and about as useful as a widely cited definition of intelligence as "that which is measured by the intelligence tests." Although Guthrie (1935) implied that his subject was learning, he too was concerned with the control and manipulation of behavior. For him, learning was a matter of what you did the last time. Skinner (1938), although he too was considered a learning psychologist and, indeed, contributed to the field of education, was primarily concerned with controlling behavior. One searches his works in vain for any principles of learning - there is only the one principle of reinforcement which has primarily a motivational function and has never been shown to play a specific role in learning. It has no more or less merit than Guthrie's principle of "We do what we did the last time." Skinner even restricts the reinforcement principle by telling us we do what we did before only if it was rewarded after we did it.

One of the major concerns of textbook writers of the time was the question of how many kinds of learning there are. Hull (1935) once answered by saying that there were as many kinds as there were theorists, and in his day there were about 13 more or less prominent theorists, counting such commentators as Freud, Lewin, Kohler and Koffka. Hilgard and Marquis (1940) resolved the issue by reducing the kinds of learnings to two (1) Pavlovian conditioning, and (2) all of the rest. The nonPavlovian views could be regarded in one way or another as Thorndikian. Operant and instrumental learning became the alternative to conditioning. Various efforts were made to eliminate one or the other or to reduce the two to one without much success. Neal Miller devoted a decade of his career to demonstrating that autonomic responses (long assigned to Pavolv as his field) were subject to instrumental, that is, reward training. His efforts, however, by his own admission, were not entirely successful or satisfactory (Miller, 1972).

Despite the reams of research with the label "learning," very few psychologists actually studied learning. Most "learning psychologists" were concerned about behavior and its prediction and control. Perhaps Ebbinghaus (1885/1913), the first to study learning, was among the last to do so. He actually faced the question of how a new response comes to be made to a stimulus to which that response had never been made before. By analogy, how do you get to call someone by name after the first introduction? The difficulty faced by Ebbinghaus was that he recognized that if he used only one nonsense syllable as a stimulus and one as a response, the learning happened too quickly to be observed. He resorted to lists of syllables and found out a lot about lists but very little about how any pair of syllables could be cited in sequence. No one after him ever did either. It is proper to note that Pavlov (1927) also actually attacked the problem of how a single response is made to a single new stimulus, but he too got sidetracked by his fascination with the response and did not study the formation of the association between stimulus and response beyond some guesswork. According to the astute analysis by Mowrer (1960a), Pavlov's dogs were not learning to salivate at all. That they knew how to do or at least could do, not necessarily by trying or wanting to. What was happening and what Pavlov ignored was that the dogs were pleased (happy, hopeful) at the arrival of food and later at the appearance of the CS and that part of the pleasant feeling engendered included salivating. (My cat salivates copiously when being stroked. The relationship between pleasure and salivation could bear some research.) Pavlov's dogs were doing a lot of other things, but Pavlov preferred to measure saliva. The pre-occupation by Pavlov with saliva and the similar preoccupation of Thorndike with string-pulling and Skinner with bar-presses and pigeon pecks cannot be described as leading to a clear picture of learning. Although we are learning all through our lives, we still do not know what is happening when we learn. We might note that if we study animals learning rather dumb things like running through mazes or people learning to recite lists of nonsense sounds, we are not likely to come up with many clear notions of how a single "thing" is learned. How do we learn to call an apple an "apple"? Or Mr. Rumpelstiltskin "Mr. Rumpelstiltskin"?

2. Autoshaping and Incidental Learning

In the late sixties an interesting phenomenon was reported by Brown and Jenkins (1968). Pigeons were placed in a box with the usual target key. The key was illuminated and food was delivered without any requirement that the pigeon peck the key. In short, there was no contingency. The pigeon was not reinforced for pecking. The situation was directly Pavlovian in that a signal was presented (the lighted key or CS) and food (US) followed. In a short time the pigeon began to peck the key in the manner previously described for countless pigeons as typical "operant behavior." Even more interesting, Williams and Williams (1969) introduced what has come to be called "ommision" training. In this case the food is not presented if the pigeon pecks at the key; only if the pigeon ignores the key will the food be delivered. In such an experimental set-up the pigeons peck at the key anyhow. Here the contingency is clearly a negative one, but the key-pecking behavior follows willy-nilly. The positive behavior with a negative contingency is at least confusing but the function of reward in learning and indeed the whole question of the reality of instrumental or operant learning comes into serious doubt. Omission training calls for more study as the role of imagery here is also unclear. Are more trials needed for extinction of such a response in pigeons?

In human verbal learning situations rewards are rarely given in any concrete form. The subjects are supposed to self-reward themselves through "knowledge of results" even though they learn without such knowledge and even when they do not know they have been learning anything. Considerable learning has been demonstrated in many studies of "incidental learning." Such studies call into question not only the necessity of reinforcement but the equally important role (in instrumental learning) of the overt reponses which must be observed and reinforced. We know now that we can learn a great deal without showing any overt performance or behavior for the behaviorist to study or observe. We can learn and remember things we are not supposed to learn and are in a way forbidden to know (Bugelski, 1970). We learn a great deal *incidentally* while tying to learn or while doing other things. In the old learning laboratories where subjects hunched over nonsense syllable dispensers they did behave, to be sure - they pronounced or at least looked at the syllables. Certainly they could not learn an unexamined list. In nonverbal situations the behavior is not so clear. You watch a television show and then can describe some of it without having tried to learn, without anyone else reinforcing you, and with no great intention or incentive to remember what you see. Your behavior while watching a TV football game might consist of drinking beer or munching potato chips. Later you might be able to describe the game to some extent but not how many chips you ate or bottles of beer you consumed. The overt behavior might be the least significant aspect of the entire learning episode.

II. THE NATURE OF LEARNING

The basic issue in learning is and always has been: Given a stimulus, A, and another stimulus, B, under what conditions will stimulus A or some variation of it bring about some activity in the subject (presumably neural to begin with) that would ordinarily occur only in the presence of B? Note that "activity" here includes motor reaction as in practiced skills and "knowledge of facts" as when we are able to make statements to others or to ourselves. Ebbinghaus created difficulties for himself by adding stimuli C, D, E, etc. Pavlov wisely stuck with A and B but, as we have noted, got too involved with the overt expression of B. We have already noted that psychologists following Pavlov, Thorndike, and Watson also got too involved with the overt consequences of B. We return to Mowrer's argument. Mowrer (1960a) pointed out that Pavlov overlooked the important reaction to B (food), namely, some kind of positive emotion. Mowrer prefers to examine that first, more direct, more important reaction, the emotion (pleasure; "hope" in Mowrer's description). Again, if the nature of B is negative, for example, an electric shock, the emotional reaction (fear) might well be the most significant consequence. How that emotion (fear) is expressed might well vary with individuals and circumstances. The expression of any overt behavior might be prevented (e.g., by placing a rat in a squeeze box with no place to run) and show up only later in different situations. Mowrer began to identify the emotional reactions to positive and negative stimuli as the major part of the meaning of such stimuli. Following Osgood et al's (1958) leads relating to the evaluative factors uncovered by his semantic differential research, Mowrer now could suggest that the meaning of anything consisted largely in how one felt about it. Such a meaning, Mowrer recognized, was only partial, connotative, however significant. The problem of denotative meanings was still unsolved. What is the meaning of an apple, for example, besides the degree of like or dislike you might experience when one is encountered? Here Mowrer, to his credit, suggested that the denotative aspects of meaning might consist of any imagery that was aroused in addition to the feelings. In the now famous sentence "Tom is a thief," the meaning of the sentence would consist of how you felt about Tom and thieves, and the imagery you might have of Tom and thieves. If you don't know any Tom, the chances are that the sentence is rather meaningless. The average person's imagery of thieves might also be rather poor and consist mostly of the emotional reaction. Later on, Paivio (1971) also suggested that imagery and meaning might be closely related. Our image of a "crouching lion" might be all that is involved in the meaning of the expression. Paivio did not emphasize the emotional reactions so important to Mowrer. Niether Mowrer nor Paivio chose to spell out the specific mechanisms underlying imagery. Mowrer followed the Leuba (1940) suggestion of conditioned sensations, which will be considered shortly, while Paivio chose an approach that defined imagery strength in terms of student ratings of the relative ease of the arousal of imagery to presented word stimuli. It should be apparent that the term "imagery" has now been used in two different ways. Mowrer's use refers to sensory or perceptual (subjective) reactions. Paivio's usage is based on experimental procedure. In what follows, an effort will be made to provide a third usage which will not violate and may even support the other usages.

1. Conditioning as Imagery Formation

When two stimuli, 1 and 2 are presented more or less simultaneously or in close contiguity, it is axiomatic that they generate neural action currents that differ in some respects, in particular neurons involved, at the least. For the present purposes, there is no need to specify which particular neural units are involved after any stimulation. Hebb (1949) called such neural actions "cell assemblies." We can be satisfied with calling them N_1 and N_2. However complex the stimuli S_1 and S_2 (a bell and food, two nonsense syllables, two people seen together or in quick succession, an object and a word, two pictures), the basic assumption made by all psychologists is that N_1 and N_2 will occur given the sensory capacities of appropriate organisms. Without this assumption psychology could not prosper or, indeed, exist.

It is then necessary to make one additional assumption, also commonly made by all psychologists, starting with Aristotle, that when two different neural actions are taking place at the same time, some components of the separate neural activities might become intermingled, connected, or associated, the one with the other, that is, related in such a manner that on a future occasion, the activity of N_1 could initiate activity of N_2 to some degree and vice versa. Such a neural "model" may be objectionable to anti-reductionists, but the worth of a model is its utility, not only in practical applications but in its capacity to make intellectual sense of a variety of kinds of observations.

In the Pavlovian situation we have the basic example of pairing two stimuli. Pavlov chose to call his stimuli "conditioned" and "unconditioned" or "CS" and "US" but the practical distinction for him was that the US was almost guaranteed to elicit an overt, measurable response. It is desirable to cover the possiblility that not all responses are observable under ordinary conditions. If S_1 is the spoken word "apple" and S_2 is a real apple or picture of an apple, shown to an observer who also hears the word, we cannot observe any responses. We can say that the person heard and saw but that is about as far as we can go. The person might concur. The responses might be only the neural consequences of the stimuli. We can label such consequences N_1 and N_2. In Figure 1, we depict the Pavlovian situation without his terminology.

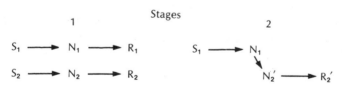

Figure 1. Pavlovian conditioning and/or the formation of imagery.

Note that if S_1 is a bell and S_2 is food and R_2 is salivation, in Stage 1 we can ignore R_1 and concentrate on R_2, if our interest is Pavlovian. Neither R_1 nor R_2 can occur without some neural action (N_1 and N_2). In Stage 2, after sufficient temporal coincidence, N_1 can initiate N_2 and lead to R_2 The prime is attached to the letters to suggest that N_2' will not be exactly like N_2 and, consequently R_2' will not be precisely like R_2 in various measurable aspects, for example, amount and latency. If, to use our earlier example, the word "apple" is heard while the subject is viewing an apple in Stage 1, again, in Stage 2, S_1 will arouse N_1 and this in turn will arouse N_2'. The subject will have a neural activity corresponding in come degree with that which follows viewing a real apple. The only difference between the function of an image and Pavlovian conditioning amounts to a concern over the overt response to N_2'. The subject, in the case of image arousal, may or may not be able to describe what is going on inside him. At the moment, we are concerned only with identifying an image as a conditioned neural reaction and a conditioned response as an overt aftermath of an image. Note that in the case where N_2' is an image, it cannot be an exact replica of N_2 because there is no S_2 to elicit it in its regular form. If N_2' occurs at all, it will probably be less complete, less prolonged, and not as "strong." In every respect it will have characteristics that psychologists since Wundt have been assigning to images.

It is obvious then that a neural activity, N_2', may be aroused normally via the presence of some other neural activity (N_1) although there may be other and multiple neural actions that might also generate N_2'. When the actual source of the neural activity that arouses N_2' cannot be identified, we can talk about "spontaneous" arousal (Hebb, 1949) although there is nothing spontaneous about it. It is just unidentified. Any new or novel "ideas" we may have could be the results of such random or indirect activation of N_2' in the presence of other stimuli previously unrelated to N_2'. When Scrooge explained his ghostly visitor as the consequence of an undigested bit of mutton, he might have been close to the truth.

Note that nothing has been said about pictures in the mind viewed through some mind's eye. Such verbiage may be misleading and shifts the topic of imagery to another level of discourse. What does need to be said is that N_1 and N_2 are neural reactions. Somehow, in early introspective psychology, such neural reactions came to be called "sensations" and came to be identified with "subjective," "mental," or "conscious" activities in some alleged "mind." Robert S. Woodworth (1938) had made a futile attempt to define sensations as the "first response of the brain to a stimulus," a definition without subjective reference, but his effort fell on deaf ears, as well might this one.

2. Images as Conditioned Sensations

Clarence Leuba in 1940, on the basis of reports and behavior of hypnotized subjects, described images as "conditioned sensations." Had he followed Woodworth's lead, he could have said "conditioned responses" but, again, had he done so, he might have been accused of confusing subjective experiences with conditioned responses and conditioned responses, as everyone knew, only happened in dogs. But, the point is that images are conditioned reactions and moreover, all conditioning amounts to the formation of images. When a Pavlovian dog hears a bell, a neural response (an image) corresponding to the sensory features of food will also occur. Along with the imagery, there will also be conditioned emotional reactions from prior experiences with food. Whether the dog has any subjective experiences is not the question. Only people can talk about subjective experiences and so the issue resolves itself into the question of whether anyone can talk about imagery, that is, describe some neural action as it is occurring or after it has occurred. The neural activity may be so rapid and brief and quickly replaced that it might escape any chance of being detected or described by a human subject. "Imageless thought" might not have been so imageless as reported by the Wurzburg school. We will address the subjective aspects of imagery later. For the moment we have identified images as conditioned responses in the brain and conditioning itself as the formation of associations between neural reactions. Whether a modern audience finds this acceptable or not, it more than likely would have been acceptable to John B. Watson (1914). Had he chosen to define sensations as responses, as did Woodworth, psychology today might be a strikingly different enterprise. His premature rejection of images as the "ghosts of sensation" stopped consideration of imagery for some 40 years. The legacy of Watson still makes imagery suspicious and unpalatable for some psychologists who, though driven to recognize some kind of internal activity as a basic feature of all behavior, prefer to deal with computer simulations and endow them with "representations," "propositions," and other imaginary mechanisms. Some psychologists prefer these terms because they seem to be more "neutral." I fear they will not remain "neutral" for long. Their neutrality itself is questionable as each has its own connotations.

The reason for emphasizing the terminological distinction between sensations and responses is not merely a matter of literary preference or taste. Sensations have always been treated as some kind of simple entities, consequences of brain stimulation, to be sure, but somehow impotent in themselves; combined with "past experience" (how or by whom the combining is done was never considered), they produce "perceptions." They can be talked about or rated as in psychophysical research, but in themselves they appear to have no direct consequences. Responses or reactions of a motor nature, at least since John Dewey's famous paper on the reflex arc (1896), have had consequences - kinesthetic feedback - or as

Guthrie (1935) called them, "movement produced siimuli." Such new stimuli, neural events produced by movement, were now deemed to be effective in producing new reactions which, in turn, produced new stimuli, and so on. Dewey should have realized that auditory and visual stimuli also produce consequences and that the consequences do not cease the moment the stimuli are withdrawn. Hebb (1949) did recognize and emphasize the continuing reverberating, perseverating consequences of sensory stimuli of any type, particularly their capacity to initiate activity in other "cell assemblies." With such perservative after-effects of sensation-produced stimuli, we have a potential answer to the question of the initiation of behavior or voluntary action.

3. Sensory-Sensory Conditioning

In 1939 W.J. Brogden reported a novel experiment. Dogs were stimulated with 200 pairings of a light and a tone with no contingent requirements and no concern over any overt responses. After such "pre-training" the dogs were subjected to shocks following presentation of one of the former stimuli (e.g., light and shock) until they made reliable avoidance responses. At this point they would then be tested with the other stimulus to see if the earlier pairing had any effect. Brogden was able to show that in this and subsequent experiments the dogs reacted to the tone as if it were the light and vice versa, in contrast to control dogs who had been subjected to the same number of lights and tones presented separately. Following Brogden's lead, there was a flurry of research in what Brogden had labeled "sensory pre-conditioning" (SPC) as if it were something that somehow differed from conditioning proper. Seidel (1959) reviewed a large number of such studies covering work with humans as well as animals and concluded: "At this point, SPC seems generally substantiated as a phenomenon in learning. Further, there are indications that the required conditions for its occurrence seem to be little more than repeated stimulus contiguity." Birch and Bitterman (1949) had earlier concluded that "The results of the sensory pre-conditioning experiments require us to postulate a process of afferent modification (sensory integration...."

Whatever "sensory integration" or "afferent modification" meant to Birch and Bitterman, we can note that it referred to the experimental pairing of two stimuli with no regard to any contingencies of response or reinforcement and the subsequent finding that one stimulus could substitute for another to some degree. In the present view, this finding provides an experimental base for the interpretation of imagery. It is unfortunate that in the period of the forties the leading theorists ignored the sensory-sensory conditioning studies, at least officially, and the work of Brodgen and his followers has been largely forgotten.

4. Synesthesia

The alleged experience of synesthesia has, like sensory-sensory conditioning, eluded serious attention from researchers. The positive findings of Ellson (1941) were challenged by the negative findings of Kelly (1934) and research by learning psychologists has lagged since then. It may be that the difficulties in arranging appropriate experimental settings have discouraged research, but the occasional subjective reports of such experiences call for some definitive study. While positive findings would be welcome, they are not required for the present model.

5. The Conditions Required for Conditioning

In summary, so far, it is asserted that Pavlovian conditioning amounts to pairing neural activities from stimuli 1 and 2 with the resulting association of the neural activities under some "conditions." The important conditions amount to two:

(1) There must be sufficient time for the two neural activities to generate some interconnections. (Pavlov achieved this by repetition but repetition was about the only way he could attain his results and test for the occurrence of conditioning. Prolonging the ringing with the eating might have been distracting although possibly more effective.) The point is, of course, that the time of exposure to the stimuli must be adequate. How much time any particular task or exercise requires is not known until appropriately tested. Paired-associate nonsense syllable learning has been shown to be a matter of the total time spent (Bugelski, 1962) rather than of the number of trials. Too much time is just as wasteful as too little time. For good imagery associations one appears to need from 5 to 8 seconds and sometimes longer (Bugelski, Kidd, & Segmen, 1968). I have observed children taking 30 to 40 seconds trying to pair images of "a bun and a kite." Any single image can occur virtually instantly. It takes some time for two to be associated. Subjects wait for "suitable" imagery to occur and do not signal readiness to proceed until they feel satisfied. The time required to master various skills should also be considered. Every athletic skill consists of many and uncounted separate subskills. It takes years to become a tournament style tennis player. Grand masters in chess and piano playing also spend years at their boards. Olympic champions for the 2000 AD games are already beginning to train and unless they got started by the age of 2 or 4 they will have little chance to perform. Time is truly of the essence.

(2) The second condition is that there should be no distractions or

interference. The time-honored practice of exposing nonsense syllables for 3 seconds is probably the least efficient procedure for learning for both reasons in that there may be insufficient time for any pair to become associated and other syllables keep coming up so that many undesired associates are created (Hull, 1935).

Under normal human learning conditions as in the classrooms, for example, there are multiple sources of distractions and interference. The very presence of other people may be a source of much interference for some learners. The isolation associated with teaching machines may be a strong factor in their effectiveness. Of course, seriously concerned adults acquire techniques by which they can tolerate various sources of distraction. Some can even learn to be deaf when television commercials appear on their screens. We can and do learn under many distracting conditions, but the efficiency of the learning must be questioned.

In recent years there has been much discussion of what might amount to a third condition for learning, namely, what has been called "processing" by Craik and Lockhart (1972). When a learner is in a situation where he has been asked to deal with some form of stimulation, he is said to be "processing" the material. He may process "deeply" or at some other level. In terms of a list of words, for example, the learner might just scan the words (superficial processing); he could vocalize or repeat the words as rapidly as possible before succeeding words appear; he might think of rhymes, meanings, form sentences, try to create images, etc. With each level or kind of processing there might be a different degree of efficiency in later retention tasks. Rather than talk about the learner as "processing" something, it might be more accurate to describe the situation as one where a learner with certain earlier learned habits of approaching learning tasks has reactions occur to and within himself that help or hinder the formation of associations. Thus, if he has learned to attack what he regards as a rote learning task by repetition of the items, he will start to recite or repeat the items. If he has learned or been instructed to form sentences, he might try to follow such instructions. The materials and the situation, including instructions, may process the subject as much as vice versa. By so-called deep processing the associations that do occur to the subject (and no one can determine which associations will occur - they either occur or do not) may be less subject to interference from other materials or stimuli; they might even be facilitated by activating prior learnings which might then assist in the recall (see Ahsen, 1979). Some kinds of processing amount to changing the nature of the learning task. The test may remain the same, for example, number of words recalled, but what was learned might be very different. Until we learn more about processing and what is actually going on in the subjects when they undertake a learning task, it might be wiser to think of different ways of processing as different ways of isolating the new associations from interferences. We may be able to rest with only two basic conditions for learning to occur: contiguity over time and freedom from interference.

There appears to be no real dispute about the facts of conditioning as thus described. We know stimuli cannot be associated - learning does not go on in the external world - it goes on in the head of the observer. Nor are stimuli associated with responses, except indirectly. Only neural reactions can be associated and, when they are, it does not matter what the external stimuli were or even if there were any.

It is concluded that responses (behavior, performance) are a matter of proper stimulation leading to natural physiological consequences. For any overt action there has to be a motor neural impulse. In the conditioning laboratory that motor impulse is generated by S_2 but, as a function of pairing S_2 with S_1, it can be indirectly initiated by the N_1 in part, if not completely.

III. VOLUNTARY BEHAVIOR

One of the characteristic differences between reflex action and voluntary responses is the typically longer latency of the voluntary behavior. Voluntary eyelid blinks to a signal are about three times longer in latency than reflex blinks. Voluntary blinks are difficult to distinguish from conditioned blinks (Peak, 1933). It appears that voluntary behavior is less directly activated than reflex reactions. The fact that instrumental behavior is rather casually identified with voluntary action suggests that an adequate account of either would amount to an explanation of both. Psychologists concerned with operant behavior or instrumental behavior may have chosen these terms to avoid coming to terms with volition. Such euphemisms may prove unnecessary.

How is voluntary behavior to be explained?

1. Instrumental Conditioning and Imagery

Proponents of operant or instrumental behavior do not deny Pavlovian conditioning; they merely relegate it to a somewhat less dignified or "autonomic" realm. It is all right for emotional reactions, PGR's, salivation, etc., but inadequate as an explanation for target pecking, bar-pressing, speaking, reading, and the like. Here responses are emitted and, if reinforced, continue to be emitted. Reinforcement is the key, and, of course, it works as it always has in all of evolutionary history. But, it is argued here that reinforcement is nothing more than the stimulation of positive emotional responses commonly referred to as pleasure and satisfaction. It is, if anything, a motivational operation, not a learning one. McCullogh (1983), in commenting on this topic, pointed out that reinforced responses will be repeated - and they will, of course - but it is the repetition that strengthens the learning and not the reinforcement. The learner also learns that reinforcement will follow. Workers in behavior therapy introduce reinforcement to "modify behavior" and then withdraw it to demonstrate that it has been the effective agent in the changed behavior. Has anyone ever questioned or denied such an effect? Of course, behavior is controlled by reinforcement (both negative and positive). School children stay in their seats on days when valued tokens are provided. When the tokens are withdrawn, they return to old seat-leaving behavior. Did they have to learn how to sit? Did they forget how to sit when tokens were not provided? Do they relearn to sit when tokens are reintroduced? Certainly we know better.

The basic question in operant situations is not then one of learning how to peck a target or press a bar, but why the action takes place at all. Pigeons know how to peck before they ever are placed in boxes with targets. They peck at anything that attracts them. Assuming no shaping, we can accept the first peck or press as an accident. But why does a pigeon peck repeatedly or a rat press a bar time after time, even if the food dispensing mechanisms jam occassionally and there is no reinforcement? Partial reinforcement is not an explanation. It is a description of a procedure.

2. The Initiation of Action: Instrumental Learning

In his analysis of learning and symbolic processes, Mowrer (1960b) faced the question of the initiation of action, the crucial issue in the instrumental or operant situation. When Guthrie (1946) accused Tolman of leaving a rat in the middle of a maze "lost in thought" he believed he had pinpointed the weakness of instrumental theory. In his own solution Guthrie had no problem. Stimuli were conditioned to responses and, given the appropriate stimulus, the rat would move as it did the last time that stimulus complex was present. This simple answer did not result in attracting many endorsers or acceptance, and the issue remained unresolved. Mowrer did ask the question but found no answer. Skinner's (1938) operant view has the organism emitting response by the thousands, but the basis for the emissions is not explained by Skinner any better than by Guthrie. To say that an organism responds in some way because of prior reinforcement may have some indirect relevance to the matter when it does actually respond and when it has been reinforced, but the findings mentioned earlier of incidental learning and autoshaping raise some serious questions. The learning and performance distinction highlighted by all of the incidental learning research remains as a crucial issue. In a moment the issue of the initiation of action will be addressed. It will be argued that the question of voluntary behavior can be approached perhaps only partially at present, by rejecting the concept of operant or instrumental learning and by accepting a straightforward conditioning model where the conditioning is presumed to be, as Pavlov originally asserted, between the neural consequences of two separate stimulus inputs, the CS and US or S_1 and S_2. Whether the presentation of the S_1 or S_2 is followed by some overt responses or not must be considered a function of many other possible variables. Dogs do not always eat presented food. In order that eating should occur at all, there must be some degree of hunger, no illness, absence of fear and distracting or novel stimuli. There must also be some familiarity with the food substance. In the Skinner box the rat does not behave like an eating-pressing machine. There are moments when it takes "time out" to scratch a stray itch or perks up its ears to some, quite possibly, "imaginary" sound.

3. Ideomotor Action

In 1890 (before Pavlov's findings) William James proposed his theory of ideomotor action. It was, in the style of the times, cast in mentalistic terms and amounted to a statement that if you thought of doing something, you would have some tendency to do it. While a wish might be father to a thought, the thought is father to the act. If you think of your finger tingling, the finger will tingle, said James. Of course, if it does not tingle, you must not have thought strongly or directly enough or you thought of it not tingling. Similarly for any action. If you think of doing something the thought will, somehow, initiate the action. What a thought is and how it arises or how it could have an effect on action was not made clear even though the matter was settled in James' view. In his day James could go no further than this statement. When Pavlov arrived on the scene no one saw the relevance of conditioning to ideomotor action until some seven decades later when Greenwald (1970) took into account the messages of Dewey, Guthrie's movement-produced stimuli, and Hull's emphasis on the antedating feature of conditioning. It should be noted that such antedating is the true hallmark of learning. One must react before the US is presented if learning is to be demonstrated in any situation. In human terms one must give the answer before a prompt is provided to demonstrate knowledge.

But, the antedating response that appears to a CS might not be a total replica of the response to the US. Sometimes it might be only partial or fractional. As noted earlier it might not be overt, that is, observable, at all and yet a response might have occurred. If a dog does salivate to a bell, the salivation would be such a fractional part of the response of eating. Because the fractional response occurs before the US it can be called an antedating response. Hull (1943) called such a "fractional antedating goal response" an "r_g" or "Little Argy," and regarded them as the "surrogates of thought." If we now recognize that these r_g's had to be initiated by neural activity related to the conditioned stimuli (S_1) we can describe the rg's as the physical expression of images which precede them. Hull need not have required an overt r_g in order that a thought should occur. The images might occur without a physical expression for a variety of reasons. A satiated dog might well have an image of food when the bell rings but be in no mood to salivate. My cat, comfortably at rest, does not always run to the kitchen when the can opener is operating, although she usually does. She will, however, open at least one eye.

We have now identified thoughts with images. Perhaps this will be more acceptable than identifying them with r_g's. I will not pursue the issue of "abstract" thoughts in this paper as I have little confidence in their existence. A colleague of mine insists that he has abstract thoughts of such matters as a "rigorous editorial policy" but I choose to regard his statement as verbal shorthand for some rather complex imagery. While this claim might appear rather cavalier, I prefer to regard it as a modest admission that I do not know what an abstract thought is or could be. I shall return to the issue of antedating conditioned responses and their consequences.

In 1970 A.G. Greenwald suggested that just as the conditioned response comes forward in time, so do the proprioceptive stimuli generated by the response. Remember that we are dealing with neural activities, and it is highly plausible that the neural activity we call the kinesthetic stimulus could very easily begin to come into action and become conditioned to the CS, just as the neural action related to the US is conditioned to the neural action of the CS. With repetition it is possible for the kinesthetic stimuli to come into play along with the response itself, and, eventually, even prior to that response (see Figure 2). Given such a situation, the kinesthetic stimuli could, in effect, initiate the response which originally gave rise to them. Because such kinesthetic neural activity could also be conditioned to other stimuli, such other stimuli could then also become capable of generating the response by activating or generating the kinesthetic stimuli or even more directly. Here we have the thought becoming a legitimate parent of the act. The original response which was the cause of the move-ment-produced-stimuli has now become the effect of such stimuli (see Figure 3).

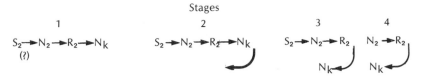

Figure 2. Voluntary action. Modified from Greenwald (1970). In Stage 1 a stimulus which might not be known or identifiable excites a response via a motor neuron action (N) and necessarily gives rise to kinesthetic action (N_k). With repetition of the response the kinesthetic action tends to come forward (Stages 2 and 3) and become temporally contiguous with N_2 (Stages 3 and 4) with which it can then become associated. By Stage 4 it can initiate the response via N_2 if it should become active for any reason.

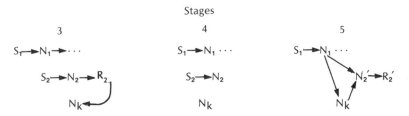

Figure 3. Instrumental learning. Starting with Stage 3 of Figure 2, external stimuli (S_1) are shown as leading to N_1 but responses to N_1 are ignored. S_2, an unknown and perhaps unidentifiable stimulus, leads to R_2 and consequent kinesthetic stimuli (N_k), which have begun to antedate R_2. In Stage 4, N_k is shown as contiguous in time with N_1 and can be associated with it. By Stage 5, N_1 can initiate N_2 via N_k and subsequently could do so in the absence of N_k through more direct association with N_2.

In the laboratory example of the Skinner box, a bar press once executed for any accidental reason could be related to the kinesthetic after-effects, and if these were conditioned to other stimuli, there would be a tendency to execute the response any time such other stimuli impinged on the organism. Such other stimuli surround the animal and can occur within it. Thus, drive stimuli could be present at the time of the original press. The visual stimuli of the box and bar and the cutaneous stimulation of the bar when it is touched would also be present, as would sounds from the food delivery mechanism. I have often observed a rat in one box run to and investigate a food cup when rats in other boxes pressed bars and produced the clicks associated with food delivery. Such investigatory responses have been conditioned to the noises involved in bar pressing (Bugelski, 1938).

It is argued here that instrumental behavior and, presumably, all voluntary action is a product of the conditioning of proprioceptive stimuli to other stimuli that are contiguous in time with the development of movement-produced-stimuli resulting from any activity. The proprioceptive stimuli then initiate the action through their association with the motor neural activity involved in the action. Once such conditioning is established, visual, auditory, or other kinds of stimuli can also become associated with the motor neural action and initiate activity without relying on the appearance of kinesthetic neural action. There is no reason to create a category of instrumental learning as if it were some other kind of learning differing from simple conditioning or the association of two neural activities.

It is not proposed that the above account of instrumental learning and voluntary action is anything but a speculative extrapolation of the analysis presented by Greenwald. It was introduced here more as an invitation to psychologists to return to what I consider the basic issue for psychology, namely, the mind-body problem. It may be that a serious examination of the nature of imagery will unravel that Gordian knot without Alexandrian measures.

IV. THE SUBJECTIVE ASPECTS OF IMAGERY

Images are commonly regarded as internal (percept-like) activities in the absence of external stimuli that might normally arouse such activities. Actually there are no external stimuli that could directly arouse precisely and exactly the kind of internal activities commonly described as images. Images, then are indirectly generated, even when and, perhaps especially, when they occur as reactions to words. Recall Figure 1. A spoken word like "apple" is and remains a noise unless it brings about something like the neural activity that occurs in connection with real apples. Recently Mary C. Potter et al. (1984) demonstrated that in bilinguals or with subjects learning a second language the words of the two languages are not directly associated with each other; rather, each word is associated with an underlying "conceptual system." I prefer to say with an image of the word referent. My study (Bugelski, 1977) of English speakers with Spanish backgrounds in childhood found that equivalent words in the separate languages referred to different physical objects albeit of the same class, for example, a *carton* of milk and a *bottle* of milk.

We have identified the image as a conditioned neural activity. The conditioning of kinesthetic stimuli to other stimulus activity as just described meets the definition of images-they share the same features and are qualitatively the same kind of neural events. We can argue that all conditioned kinesthetic activity is imagery, although not all imagery is kinesthetic activity-it can be neural activity generated by any other kind of stimulation. What remains for consideration is the alleged subjectivity of imagery. We can now turn to this question. Skinner (1953) has carefully examined the issue of subjective reports. He points out that all of us learn through social assistance to talk about such matters as bruised knees and cut fingers with solicitous remarks about "Does it hurt much?" We also find ourselves reacting to internal discomforts and fears with social suggestions about "tummy aches." Amputees talk about sensations from "phantom limbs." We all describe "pins and needles" in our legs. In time we learn to talk about some of the things that do occur in our "insides, " but never with the accuracy that we acquire in describing external events. Frequently we find our medical practitioners telling us how wrong we are about the actual source of a pain. They tell us about "referred' pains or attribute various distant symptoms to our teeth or whatnot. We are not very good about describing internal events. When it comes to kinesthetic stimuli we are largely lost and although we could not walk without them, we are rarely able even to mention them. Most people never heard of them. Athletic coaches learn to guide our golf stances, head positions, etc. to the net effect of having our future activities in those areas improve. Mowrer (1960a) has emphasized the role of kinesthesis in sports, of how basically all skills are a matter of conditioning kinesthetic stimuli to "feeling better" about certain orientations of our bodies as we are about to shoot pool or shoot baskets.

When the appropriate kinesthetic stimuli occur, the motor responses follow. "Appropriate" here refers to "feeling" right or "good." Beginners often feel right at the wrong time and make errors. With practice, more effective discriminations develop as a function of extinction. One begins to feel "wrong" when certain kinesthetic stimuli come into play and avoidance responses preclude positive action. And yet, the stimuli or actions going on inside our bodies are largely strangers to us. When it comes to describing internal "perceptual" reactions, as we do when we discuss imagery, we are rather helpless. Describing what a purse snatcher looked like becomes more than a little awkward. We do manage within some limits to use language in such cases, but the difficulties of having our internal neural activities guide our tongues or hands are more than obvious. How many of us can develop a reasonable sketch of even a seen object or person, much less the unseen and unpresent? Imagery has been commonly categorized as evanescent, incomplete, fleeting, unclear. With some simple requests like counting the corners on a block F, we are usually adequate to the task. To describe the form of a simple word like "the" would probably be beyond us. Trying to tell a modern college student how to knot a necktie might strain the best of us. In the latter instance, note how we tend to use our hands in ideomotor fashion as the words elicit associated kinesthetic imagery.

When we try to talk about imagery as "pictures in the mind," we are basically incompetent. When the "images" occur in the right hemisphere in split-brain patients, the patients may be able to react appropriately but be unable to say anything at all. All of their right-brain imagery is "unconscious." To a limited degree, such "visualizations" can be of some help in simple mnemonic tasks-in fact, in that area we, at least most of us, can become relative geniuses. Beyond such exercises, the work or role of imagery is still largely unknown. How it functions in creative thinking is a large mystery. The imagery of poets and other writers is beyond the appreciation of most of us because imagery is so personal, so individual, so tied to our own experiences that our degree of communication with others must be considered quite remarkable when we realize that we cannot actually describe our emotions and our imagery with any degree of precision. For most practical purposes a common word like "table" will have an adequate communication value. A child faced with a "Table of Contents" for the first time will be at least a little non-plused. If more than one table is present in a room and someone suggests that you put something on the table you must ask: "Which one?" By continual asking we can narrow down the "meanings" others are trying to communicate to us, but sometimes the process can take months or years, as when the Arabs propose that all peoples living in an area be allowed to live in peace. The Israelis ask: "Does 'peoples' mean nations? States? Countries?" Similar questions arise in all walks of life but, for the most part, we can manage to survive without complete understanding. Eventually, the belly ache goes away and it does not matter any more whether anyone knew how you suffered.

For purposes of getting along, we are content with the relative poverty of our imagery. If someone asks us to think of a flying white horse, we do so readily. We do get some kind of internal event running off when flying white horses are mentioned. If the white feature were not specified, it might be a horse of a different color and we can even imagine a green one. But not very well. If we were asked if anyone is riding it, we might deny that but can just as easily attach a rider, having meanwhile omitted to saddle the beast. When we describe a sore throat as "like sandpaper" we are resorting to analogies that are rather less than helpful. Probably nobody has ever had a throat that actually felt like sandpaper. It is a simile for an unpleasant condition. We are driven to the use of metaphors and similes by our inability to provide adequate details about our image activity. Such metaphors and similes are rarely accurate. Sometimes the poet uses them for other reasons than those of realistic description. When one thinks of an apple he can report that he can "almost taste it" but if an actual apple is presented it will not look like the imagined apple, nor will it taste like the alleged gustatorial image. This is not to suggest that the imaged apple did not appear real to the imager. It was a different apple. The poverty of images should not be decried. What else can one expect? Conditioned responses have always been criticized as not quite like the real unconditioned response. How could they possibly be? The proper stimulation for the real unconditioned response is a a specific form of energy or substance and if that is missing the surrogate response is almost by necessity going to be partial, inadequate, certainly not thoroughgoing.

Some people deny having any images. What they are denying is the subjective experience most of us claim to enjoy. When Watson claimed images to be delusions, Margaret Washburn (1916) suggested that such mass-scale delusions should certainly bear investigation. Her own classical investigation is not inconsistent with the discussion in this paper. It might help all of us to reread her work.

1. Voluntary Images

It has been implied in all of the above that images are not voluntary; they are reactions to prior stimulation either from conditioned stimuli, other neural processes that generate the kind of neural actions, or events that have been labeled images. A number of my critics chose to ignore this emphasis and referred to "voluntary" images. Thus, if someone is asked to image a duck or a whale smoking a cigar and the person says that he is doing so, the assertion is made that the image was voluntary. Clearly, he was not imaging the duck or whale before the request and is now not imaging something else. How such action could be termed voluntary is at best debatable if not a clear misuse of the term. It is argued here that no one can decide to think of anything or image anything out of the blue. Whatever he decides to image will already have been decided for him, presumably by his past and the present context. We image whatever occurs to us, be it from someone's suggestion or some other source. Sometimes we are lucky enough to be able to trace back through a string of associations that account for some present image which might appear rather novel; frequently enough we cannot. Hebb would find no difficulty with such a circumstance in terms of his cell assemblies.

2. Images as Epiphenomena

It is commonly asserted by those who are not sympathetic to the claim that images can function as mediators in learning and behavior that images are, at best, only epiphenomena. Psychologists and philosophers commonly take a dim view of epiphenomena, as if they were irrelevant. *The American Heritage Dictionary* defines an epiphenomenon as "a secondary phenomenon overlapping and resulting from another." The image as described here is just that, a response resulting from some prior stimulation. When people claim to be experiencing or having images, there is no point to denying their claims if subsequent behavior can be shown to be different from what it would have been if no such claim had been made or if the person involved himself correctly denied having had any image. Viewed in this fashion, there is nothing ignoble or improper about recognizing images as rather powerful, useful, and effective phenomena. Subjects in imagery studies appear to have no difficulty in acquiescing to instructions like those of Paivio (1971) to rate words on imagery scales. Nor do Kosslyn et al's (1979) subjects have trouble in imaging elephants and mice and rabbits' toes when instructed to do so. The stable and sometimes potent results of research under such instructions and the powerful effects of similar instructions in mnemonic exercises are evidence of a rather serious and important epiphenomenon.

In any causal chain of events, for example, A→B→C, there may be some "epiphenomenal" event, say E, that is consistently and reliably correlated with A or B. In such cases, especially when A or B are difficult to implement, it might be possible to introduce E and bring about the desired result. Thus, if a person agrees that he can "picture" a horse or battleship or whatever, there is no need to bring an actual horse or even picture of a horse into the situation. If the person can now behave appropriately to the epiphenomenal horse, a great deal of expense and energy can be saved. The ingenious experiments of Kosslyn et al. (1979) and Paivio (1971) on imagined distance effects related to imaged animals and the imaged rotation of solids demonstrated by Shepard and Metzler (1971) suggest that even if viewed as an epiphenomenon, the image can be a reliable research tool.

V. IMAGERY AND PSYCHOTHERAPY

Because psychotherapy amounts to a relearning program, a few remarks about psychotherapy and imagery appear to be in order. Even before cognitive psychologists began to show interest in imagery, some psychotherapists (Ahsen, 1965; Wolpe, 1958) had begun to take advantage of imagery in what appears to be a potent therapeutic procedure (see Wolpe on Ahsen, 1969, pp. 233-235). Other psychologists were not far behind, and we had the paradoxical phenomenon of Pavlovian conditioners working with what had been an epiphenomenal anathema to behaviorists for decades. But there is a major difficulty about employing imagery in psychotherapy. Psychotherapy, by its nature, relies on the use of words both by therapists and by patients. Patients must talk, and if they are to talk about their imagery, there is going to be trouble. As has been emphasized before, images are difficult to describe. The nature of that difficulty can be appreciated by recognizing the difficulty of reports of direct perception. When two people are asked to describe a littered table directly in view, the descriptions will not be identical, perhaps not even close. They will not begin and end in the same way, nor will the descriptions be equally detailed. One description might amount to "What a mess." A person reading the description might not be able to recognize the table when he saw it. We have all been trained to believe that words are necessarily related to some inner realities. While it is obvious that people can talk, a lot of words can be spoken without much being said. It is necessary to recognize that while there are many words available, there just are not enough words to insure that a specific image will be aroused or described when a given word is uttered. The trouble arises from the fact that any one word may have multiple imagery associates and arouse a great many different images in different contexts and settings or under different personal sets. Two papers by Ahsen (1979, 1982) involving his Triple Code Model deal with this issue in detail, discussing the relationship between images, body feelings and meaning, which includes recall and words; and they contain important theoretical frameworks concerning central imagery behavior.

Mowrer (1960b), as mentioned earlier, emphasized a distinction between denotative and connotative meanings. Words can arouse either or both. In the case of denotation a word like "knife," for example, can arouse a great variety of imagery in the same person and in different people on a rather placid denotative or indicative level with little or no affective reactions. We all have had experiences with Boy Scout knives, paring knives, bread knives, butter knives, even machetes and hunting knives and enjoy imagery of such articles without getting aroused emotionally. Such associations may be of no special significance, even with a patient who has some profound and serious stressful associations related to some specific knife. A word like "mother" may arouse any image from a virtual catalogue of images involving mothers, one's own or even that of Whistler's. The patient may be talking about "Mother," but the therapist may not really know what the patient is saying. It is only when the patient is describing imagery in the connotative sense, when imagery is associated with emotional reactions, that therapeutic progress can begin. A patient who was spanked by the father may retain some imagery of a frightening ogre because at the time of the spanking the patient was too young to appreciate the situation. Subsequent years of imagery associated with the father in other times and circumstances may obscure the ogre image and even favor other imagery to be aroused by the word "Father." In this sense, some images, like words, can be lies. The old image, if it is an important factor in the patient's maladaptive behavior, must be elicited and reconditioned with appropriate emotional reactions.

The use of imagery in therapy must be applied with care until we have a better image of the image. Theoretical speculations about the roles of imagery and words must be refined. Paivio's (1971) two-track theory or Dual Code Model may provide too much of a leading role for words, endowing them with the responsibility for carrying abstract meanings and sequential thinking. In this sense Ahsen's Triple Code Model, which adds the somatic response to Paivio's two codes, is a welcome addition. I have elsewhere (Bugelski, 1977) questioned the notion that words have any meanings other than the imagery and affect they arouse. What does "ineluctable insousciance" mean? Here we have an abstract sequence of words on which it is unlikely any two people would agree as to meaning. If we take a long document like the Constitution of the United States, we will find a lot of abstract sequences. We then appoint a Supreme Court to tell us what it means in any specific case, and that will vary from time to time.

Words, gestures, body language-these are all secondary developments in our life histories. Pavlov even described them as "the second signal system" and, although they are the primary tools of communication, it is the first signal system that really matters in children. Images come before words and, while arousable by words, images will differ from time to time and carry different meanings. Contrary to Bandura's (1977) notion that verbally based codes hold more information than do imaginal codes, it appears that we actually suffer from a paucity of words and an affluence of imagery but, unfortunately, we do not put imagery to much use.

When therapists ask subjects to engage in imagery of significant figures (Ahsen, 1968) or phobic situations, the specific imagery aroused must be relevant. Ahsen tries to get at the basic or "true" image that under- lies routine, easy, in effect, "cover-up" imagery. To the extent that this can be done, he may be working with significant clinical material. Other therapists appear to be satisfied with asking their patients to image something and presume that the request is being followed. Wolpe (1969) urges patients to experience vivid imagery on the grounds that the more vivid imagery will counter-condition more easily. What may be more important is that the appropriate imagery be isolated and identified.

The same caveats hold for the behavioral therapists who follow a re-inforcement procedure. What imagery is being reinforced is the unanswered question. Cautela (1977) assumes that covert behavior to an imagined stimulus is the same as that to an overt stimulus. It should be quite clear that people generally can discriminate between the imaginary and the real. A real caged lion is quite different from an uncaged one and an imaginary lion is unlikely to arouse headlong flight. There would be no patient in the office. Lick and Unger (1975) have cautioned therapists about the generality of office cures and we have to be careful about what we are doing, and what it is really taking place.

Other therapists dealing with daydreams and fantasies (Klinger, 1971; Meichenbaum, 1977) might well be wasting valuable therapeutic time if their patients are indulging in irrelevant fantasies. It can take a long time to discover what the daydreams are covering or hiding. I am using the term "daydream" in its common sense of a "pleasant reverie." Fantasies must be related to the maladaptive aspects of a patient's history. The last one to be allowed to pick and choose his fantasies is probably the patient.

The recent flurry of research into control of visceral activities represents another area of possible application of imagery. After reviewing the field and pointing out the difficulties, Miller (1972) suggested that some bodily activities might be subject to control by imagery manipulations. He did not provide any experimental support, but the suggestion might well be explored.

The "epiphenomenon" of imagery thus appears to have very practical clinical as well as scientific research possibilities. The fact that not all of our images are subjectively available or easily describable or that some are not precise "pictures" of some previously experienced perceptions does not impugn the value of whatever imagery we can enjoy. What needs to be done is to devise techniques for identifying and controlling specific images and more effective and different ways to employ such imagery. The development of athletic and other skills, as mentioned earlier, might be enhanced by devising special techniques for familiarizing ourselves with kinesthetic imagery. Some preliminary reports have appeared describing marked improvements in throwing darts, shooting basketball foul shots, jumping, even piano playing (see Richardson, 1969). Perhaps similar training in visual and auditory imagery could improve other skills. The use of imagery in ordinary educational settings is just beginning to develop. Even textbooks can be produced which aim at enhancing the learning of subjects like history via increased imagery value of the words used in the texts (Wharton, 1980).

All such efforts at training, at present, rely on the "picture metaphor." The "picture metaphor" can apply, if at all, only to visual imagery and tends to discourage interest in other sensory activities. With auditory, olfactory, and gustatory, to say nothing of the all-important kinesthetic imagery, we might have to use different "metaphors." People do report that they sometimes hear sounds that were not actually physically present. We need to consider the reports of schizophrenic voices, although they may be real enough to those who hear them. Who has not asked "Was that the phone?" when no phone has rung? Did the deaf Beethoven have no inkling of what his late compositions would sound like? The area of auditory imagery surely requires attention, particularly in the areas of speech and music. The demonstrations of the crippling effects on speech of delayed auditory feedback enjoyed a brief popularity among researchers a few years ago. Interest in such cybernetic effects appears to have dwindled and might well be revived through an imagery approach.

It is unfortunate that most psychologists cannot deal with neural events directly and that physiological researchers do not share the interests of those concerned with imagery. For the present we must, unhappily, deal with "metaphors" but our metaphors do not appear to be as remote as the poet's "books in babbling brooks." They refer rather to a substantial if, perhaps, inaccessible reality.

REFERENCES

Ahsen, A. (1965). *Eidetic psychotherapy*. Lahore: Nai Matbooat.

Ahsen, A. (1968). *Basic concepts in psychotherapy*. New York: Brandon House.

Ahsen, A. (1979). *An image theory of conflict*. Presented at Future seminars Meeting, Image Institute, New York.

Ahsen, A. (1982). Imagery in perceptual learning and clinical application. *Journal of Mental Imagery, 6*(2), 157-186.

Anderson, J.R., & Bower, G.H. (1973). *Human associative memory*. New York: Wiley.

Bandura, A. (1977). Self-efficacy: Toward a unifying theory of behavioral change. *Psychological Review, 84,* 191-215.

Birch, H.G., & Bitterman, M.E. (1949). Reinforcement and learning. The process of sensory integration. *Psychological Review, 56,* 292-308.

Brogden, W.J. (1939). Sensory preconditioning. *Journal of Experimental Psychology, 25,* 323-332.

Brown, P.L., & Jenkins, H.M. (1968). Auto-shaping of the pigeon's keypeck. *Journal of the Experimental Analysis of Behavior, 11,* 1-8.

Bugelski, B.R. (1938). Extinction with and without sub-goal reinforcement. *Journal of Comparative Psychology, 26,* 121-133.

Bugelski, B. R. (1962). Presentation time, total time, and mediation in paired-associated learning. *Journal of Experimental Psychology, 63,* 409-412.

Bugelski, B.R. (1970). Words and things and images. *American Psychologist, 25,* 1002-1012.

Bugelski, B.R. (1977). Imagery and verbal behavior. *Journal of Mental Imagery, 1, 39-52.*

Bugelski, B.R. (1983). Imagery and thinking. In A.A. Sheikh Ed.), *Imagery: Current theory, research and application.* New York: Wiley.

Bugelski, B.R., Kidd, E., & Segman, J. (1968). Images as mediators in one-trial paired associated learning. *Journal of Experimental Psychology, 76,* 69-77.

Cautela, J.R. (1977). Covert conditioning: Assumptions and procedures. *Journal of Mental Imagery, 1,* 53-65.

Craik, F.I.M., & Lockhart, R.S. (1972). Levels of processing. *Journal of Verbal Learning and Behavior, 11,* 671-684.

Dewey, J. (1896). The reflex arc concept in psychology. *Psychological Review, 3,* 357-370.

Ebbinghaus, H. von (1913). *On memory.* New York: Teachers College, Columbia University. (Original work published 1885).

Ellson, D. (1941). Hallucinations produced by sensory conditioning. *Journal of Experimental Psychology, 28,* 1-20.

Greenwald, A.G. (1970). Sensory feedback mechanisms in performance control. *Psychological Review, 77,* 73-99.

Guthrie, E.R. (1935). *The psychology of learning.* New York: Harper.

Guthrie, E.R. (1946) Psychological facts and psychological theory. *Psychological Bulletin, 43,* 1-20.

Hebb, D.O. (1949). *The organization of behavior.* New York: Wiley.

Hilgard, E.R. (1956). *Theories of learning* (2nd ed.). New York: Appleton-Century-Crofts.

Hilgard, E.R., & Marquis, D.G. (1940). *Conditioning and learning.* New York: Appleton-Century-Crofts.

Hull, C.L. (1935). The conflicting psychologies of learning-a way out. *Psychological Review, 42,* 491-516.

Hull, C.L. (1943). *The principles of behavior.* New York: Appleton-Century-Crofts.

Hull, C.L. (1951). *Essentials of behavior.* New Haven: Yale University Press.

James, W. (1890). *The principles of psychology.* New York: Holt.

Kelly, E.L. (1934). An experimental attempt to produce artificial chromaesthesia by the technique of the conditioned response. *Journal of Experimental Psychology, 17,* 315-341.

King, D.L. (1979). *Conditioning: An image approach.* New York: Gardner Press.

Klinger, E. (1971). *The structure and functioning of fantasy.* New York: Wiley.

Kosslyn, S.M., & Pomerantz, J.R. (1977). Imagery, propositions, and the form of internal representation. *Cognitive psychology, 9,* 52-76.

Kosslyn, S.M., Pinker, S., Smith, G.E., & Shwartz, S.D. (1979). On the demystification of mental imagery. *The Behavioral and Brain Sciences, 2,* 535-582.

Leuba, C. (1940). Images as conditioned sensations. *Journal of Experimental Psychology, 26,* 345-351.

Lick, J., & Unger, T. (1975) External validity of laboratory fear assessment: Implications from two case studies. *Journal of Consulting and Clinical Psychology, 43,* 864-866.

McCullough, L. (1982) I image, therefore I learn. *Journal of Mental Imagery, 6,* 44-47.

Meichenbaum, D. (1977). *Cognitive-behavior modification: An integrative approach.* New York: Plenum.

Miller, N.E. (1972). Interactions between learned and physical factors in mental illness. In D. Shapiro et al. (Eds.), *Biofeedback and self-control.* Chicago: Aldine.

Mowrer, O.H. (1960a). *Learning theory and behavior.* New York: Wiley.

Mowrer, O. H. (1960b). *Learning theory and the symbolic processes.* New York: Wiley.

Osgood, C.E., Souci, G.J. & Tannenbaum, P.H. (1958). *The measurement of meaning.* Urbana: University of Illinois Press.

Paivio, A. (1971). *Imagery and verbal processes.* New York: Holt, Rinehart, and Winston.

Pavlov, I.P. (1927). *Conditioned reflexes.* Oxford: Humphrey Milford.

Peak, H. (1933). An evaluation of the concepts of reflex and voluntary action. *Psychological Review, 40,* 71-89.

Potter, M.C., Kwok-Fai So, VonEckhardt, & Feldman, L.B. (1984) Lexical and conceptual representation in beginning and proficient bilinguals. *Journal of Verbal Learning and Verbal Behavior, 23,* 23-38.

Pylyshyn, Z. (1973). What the mind's eye tells the mind's brain. *Psychological Bulletin, 80,* 1-22.

Richardson, A. (1969). *Mental imagery.* London: Routledge and Kegan Paul.

Seidel, R.J. (1959). A review of sensory preconditioning. *Psychological Bulletin, 56,* 58-73.

Shepard, R. N., & Metzler, J. (1971). Mental rotation of three-dimensional objects. *Science, 171,* 701-703.

Skinner, B.F. (1938). *The behavior of organisms.* New York: Appleton-Century-Crofts.

Skinner, B.F. (1953). *Science and human behavior.* New York: Appleton-Century-Crofts.

Thorndike, E.L. (1932). *The fundamentals of learning.* New York: Teachers College, Columbia University.

Tolman, E.C. (1932). *Purposive behavior in animals and men.* New York: Appleton-Century-Crofts.

Washburn, M. (1916). *Movement and mental imagery.* Boston: Houghton Mifflin.

Watson, J.B. (1914). *Behavior.* New York: Henry Holt.

Wharton, W.P. (1980). Higher imagery words and the readability of college history texts. *Journal of Mental Imagery, 4,* 129-147.

Williams, D.R., & Williams, H. (1969). Auto-Maintenance in the pigeon: Sustained pecking despite contingent non-reinforcement. *Journal of the Experimental Analysis of Behavior, 12,* 511-520.

Wolpe, J. (1958). *Psychotherapy as reciprocal inhibition.* Stanford: Stanford University Press.

Wolpe, J. (1969). *The practice of behavior therapy.* New York: Pergamon Press.

Woodworth, R.S. (1938). *Expereimental psychology.* New York: Holt.

CHAPTER **5**

Computational Theories of Image Generation

Stephen M. Kosslyn, James D. Roth and Eliott Mordkowitz

One of the most striking characteristics of visual mental images is that we do not have them all the time. Images seem to come into mind when we need them, and lie in wait thereafter for future reawakening when appropriate. Simple introspection tells us that an object or scene is present in an image, but cannot tell us how images are created. Given the importance of the image generation process, upon which all other imagery

Preparation of this chapter was supported by NIMH grant MH 39478-01 and ONR contract N00014-85-K-0095 awarded to the first author.

processes ultimately depend, it is surprising that there are only the sketchiest of theories of image generation.

Perhaps the single major contribution of contemporary cognitive psychology is that it breaks into parts phenomena that were treated previously as undifferentiated and unitary. For example, "memory" is no longer treated as a single "faculty," but instead is seen as a number of processes (encoding, search, comparison, etc.) that operate on distinct structures (short-term buffers, long-term networks, etc.). Contemporary theories often specify a set of distinct processing modules, "black boxes" that operate on input in specific ways. In this chapter we will further develop the theory of the processing modules used in image generation that was proposed by Kosslyn and Shwartz (see Kosslyn, 1980).

I. CONSTRAINTS ON THEORIES

Two sorts of constraints must be respected when one posits theories of mental events. First, theories must be consistent with the empirical results. That is, the theory must be able to provide accounts for existing data and successfully predict new data. Second, theories of mental phenomena must be consistent with computational constraints, that is, with the requirements imposed by the problem itself. Given a goal, certain ends must be accomplished on the basis of specific input. In this section we review both sorts of constraints and use them to provide the initial motivation for a theory of mental image generation.

1. Empirical Constraints

Research on mental image generation has provided some five classes of results, all of which must be respected when formulating a theory.

Number of Parts

The time to form an image increases with each additional part included on the object; the time usually increases linearly with increasing numbers of parts (Beech & Allport, 1978; Kosslyn, Reiser, Farah & Fliegel, 1983; McGlynn, Hofius, & Watulak, 1974; Paivio, 1975; Pear & Cohen, 1971). This result has been found when "parts" are defined by various Gestalt grouping principles (similarity, proximity, continuity) and by presenting material sequentially and asking subjects to "fuse" parts into a composite image.

Description of Structure

Generation times also increase when more parts are present when "parts" are defined conceptually. In one experiment, Kosslyn et al. (1983)

asked subjects to "see" a pattern in a specific way; for example, the Star of David as two overlapping triangles or a hexagon with six small triangles. The time to form the image was highly correlated with the number of parts predicated in the description, even though the physical stimulus was the same. In another experiment (see Kosslyn, 1980), subjects were shown arrays of upper case letters, and later were told that the array was "three rows of six" or "six columns of three." Even though the number of letters was the same in both cases and the description was given after the array was removed, time to generate the image subsequently depended on its description, with more time being taken when the "six columns of three" description (which predicated more units) was used. This result indicates that descriptons can be stored and used in the generation process.

Description of Relations

Beech and Allport (1978) showed that people can use descriptions of scenes to form composite images of objects arranged in specific ways (e.g., with objects in different "pigeonholes" of an imaginary grid). In their experiments, image generation time increased with increasing numbers of objects in the scene. (Incidentally, in one experiment more time was taken to image a scene than to comprehend its description, suggesting that image generation is not a necessary component of comprehending such descriptions; see also Moore, 1915). Kosslyn et al. (1983) also showed that descriptions can be used to arrange images of drawings into a scene. They found that after the image was formed, more time was taken to scan between objects that should have been further apart in the image, validating the claim that the description was actually used.

Weber, Kelley and Little (1972), using a selective interference task similar to that of Brooks (1969), showed that people can use implicit speech to help them image a sequence of letters. In their experiments, people apparently silently said the names of letters of the alphabet as prompts when the letters were imaged in order and then classified them according to their relative height (e.g., "a" is short, "b" is tall, etc.). Although such verbal prompts were used to image isolated letters of the alphabet in sequence, they were not used when words were imaged.

Size

Subjectively large images of objects ("subjectively" because the size is as it appears to the subject) require more time to form than subjectively smaller images. Why might this be? One answer is implicit in the finding that more time is taken to "see" parts of objects imaged at smaller sizes (Kosslyn, 1975). If parts are added to an image one at a time during the generation process, then it will be more difficult to "see" where parts belong on objects imaged at smaller sizes. Thus, fewer parts may be added to smaller images, resulting in faster generation times. This idea was tested by Farah and Kosslyn (1981), who varied both the size and complexity of to-be-imaged drawings. They found that when people were asked to image the drawings, there was a greater effect of size on generation time for the complex drawings, which was exactly as expected if the effects of size reflect the amount of detail that is added to an image. That is, because there were more potential details to be added, the effects of leaving them off at a small size were more dramatic for the complex drawings. This finding was in contrast with that for a second group that was highly motivated to include all of the details at both sizes; this group showed the longest times for complex drawings imaged at a small size, presumably because it was very difficult to try to add parts to the small image.

Similarity

The effects of size on image formation time suggest that the more difficult it is to "see" where parts belong, the harder it is to form the image. This inference receives support from an experiment reported by Kosslyn et al. (1983). They found that patterns of similar letters (e.g., MNMNMN) required more time to imagine than arrays composed of dissimilar letters (e.g., MGMGMG), even though each letter appeared equally often in "similar" and "dissimilar" arrays. Presumably, in the alternating pattern of similar letters it was more difficult to "see" whether letters were correctly arranged as one was adding new letters to the image.

2. Computational Constraints

Given these foregoing empirical findings, computational considerations provide additional constraints for theorizing. A concrete way of thinking about computational constraints is to think about the constraints on building a computer simulation model of processing. That is, when formulating theories of information processing, it is often useful to think about what would be necessary to program a computer to mimic the observed effects. Such considerations place three main requirements on a theory of image generation.

First, the mere fact that people can generate images at all implies that there must be a processing module that can activate the information about an object's or part's appearance stored in long-term memory.

Second, the fact that people can use descriptions to arrange separate parts or objects into a single imaged scene requires a processing module that interprets the relations (e.g., "six feet above"). This module must also be able to affect the first one, so that objects or parts can be imaged in the correct relative locations.

Third, the fact that people can image multi-part objects of different sizes and locations implies that the parts are not remembered as being at absolute locations in space, but instead are encoded vis-a-vis one another. If so, then there must be a way of locating where a part should go relative to what is already in the image. Hence, we assume there is a processing module that is used to locate objects or parts already in the image. Such a module is particularly important when one is composing a novel arrangement of previously seen objects.

II. A THEORY OF IMAGE GENERATION

Theories of information processing can be cast at numerous "levels of analysis" (see Marr, 1982). Two levels have proved particularly useful for theories of mental phenomena: theories of "the computations" used, and theories of the "algorithms" that carry out the computations.

1. Computations

The Kosslyn and Shwartz (1977; 1978) theory posits that three processing modules are used in image generation. Each module carries out a set of related "computations" which work together to transform the input in some way. The modular breakdown posited by the theory was motivated in a straightforward way by the empirical findings and the computational constraints.

The first module is called the PICTURE processing module. This module is assumed to be used iteratively when an image of an object is constructed from more than one stored part; in this case, the parts are generated one at a time. This assumption allows us to explain why the time to generate an image increases as increasing numbers of parts are used to compose an object or scene.

The second module is called the PUT processing module. This module looks up the relationship between a pair of parts (e.g., "under rear wheelwell" for a car's rear tire), and uses this information to calibrate one part (e.g., a tire) relative to another (the wheelwell). In order to perform these functions, however, the PUT processing module first must locate where one object or part is in the image before it can place the next in the correct relative location.

Thus, the PUT module makes use of the third module, which Kosslyn and Shwartz call the FIND processing module. This module is a pattern recognizer, which is used to identify spatial patterns as depicting specific parts or objects. The FIND processing module is in fact "the mind's eye," looking for patterns by testing for the presence of sets of particular simple patterns (e.g., straight lines, angles, etc.). This module can only classify patterns correctly when their shapes are discriminable; thus, if objects or parts are imaged too small, the FIND processing module will fail to "see" them (at least on the first attempt; more exacting tests can then be used which will locate the parts, but take longer to do so). This module is presumably used not only to locate where to-be-imagined parts should be placed relative to previously-imaged parts ("foundation" parts), but also when one "inspects" an object in an image for a specific property, or even when one inspects a physically-present object during perception. (There are numerous findings that support the claim that the same visual inspection processes are used in perception and imagery; for reviews see Finke, 1980; Finke & Shepard, in press; Kosslyn, 1980, 1983; Podgorny & Shepard, 1978; Segal, 1971).

To summarize, the PICTURE processing module activates stored visual information; the PUT processing module looks up and interprets a stored description of the relationship among parts (e.g., "the front wheel goes in the front wheelwell"), and invokes the FIND processing module; the FIND module locates the part to which the to-be-imaged part is relative; and then the PUT processing module uses the information from the FIND module in conjunction with the description of the relation (e.g., "under") to set the PICTURE processing module. The result is that the two parts are correctly arranged in the image.

2. An Algorithm

An algorithm is an explicit set of steps that will produce a specific outcome given a specified input. It is difficult to make many predictions of patterns and magnitudes of response times, errors, or the like without a theory of algorithm because the algorithm allows one to know what will actually happen internally when a task is performed. Algorithms can be specified at different levels of detail, from a very coarse level that specifies only the order in which modules are used, to a very detailed level that specifies the steps that carry out the job of individual modules. Depending on the level of specificity, predictions will range in precision from relatively coarse (e.g., relative orderings of response times) to a very specific (e.g., exact quantities of times in different conditions). The algorithm we posit is at the coarser end of the spectrum, as is illustrated below.

The computer simulation developed by Kosslyn and Shwartz (1977; 1978) interacts with a user, who stands in for the non-imagery components of the cognitive system. The user asks the simulation to generate an image of an object, for example, a chair. The first thing the simulation does is to check whether it knows what chairs look like. In the model, each object has a file containing a list of propositions, one of which may have the form CHAIR.IMG. This is the name of the "image file" for chair. Having this name, the PUT module now locates that file and evokes the PICTURE module to form an image of its contents. Forming an image in the module consists of filling in selected cells of matrix; the matrix functions as a visual short-term memory structure, and the pattern of points within it represents the imaged object. The matrix has limited grain, so that parts of objects will blur when the object is imaged at a small size. In the model, the contents of the long-term memory image file are a list of coordinates indicating where points should be placed to depict the desired object. Obviously, the brain does not literally store lists of coordinates, and does not contain an actual matrix; the theory posits, however, that the brain *functions as if* it had these properties.

If the task requires including additional parts, then the image will be generated including details stored in separate memory encodings. In the simulation model, the user indicates whether a skeletal image (i.e., a low-resolution image of the overall form) or a detailed image is required. If a skeletal image is sufficient, as it would be if the task were to decide whether a chair is taller than it is wide, then the image generation sequence stops after the image associated with the object as a whole is generated. However, if precision is required, as would occur if one were asked if the arms are longer than the legs, then a detailed image is generated.

The PUT processing module orchestrates the addition of details to the image. It first looks in the object's propositional file for entries of the form HASA.CUSHION. These entries indicate parts of the object. If a part is listed, the propositional file for that part is looked up. The PUT module then looks for the name of an image file listed in this file, for example CUSHION.IMG. This indicates that one knows what the part looks like; if one doesn't know what the part looks like, there is no sense in going further. If such an entry is present, the PUT module now proceeds to generate the image.

The next thing the PUT module does is to find the "location relation" for the part. For example, for cushion the relation is "FLUSH-ON.SEAT." The relation then must be broken into two parts, the relation proper, "FLUSHON," and the "foundation part," in this case "SEAT." The relation indicates how the part should be placed with respect to the foundation part.

The PUT module next evokes the FIND module to locate the foundation part in the image. This module is given the name of the to-be-located foundation part, "SEAT." The first thing the FIND module does is look up the propositional file for SEAT and locate a description of the part's appearance. In the model, the description is a sequence of numbers. Each number corresponds to a test (e.g., "find the lower right-hand corner, trace up until hitting an intersection"). The FIND module executes these tests in order on the image. If all of the tests can be performed successfully in sequence, then the part has been located. If it cannot (e.g., because the object is so small that parts are obscured by the "grain" of the array), then it so indicates.

If the part has been successfully located, the FIND module indicates the part's location and size to the PUT module. The PUT module then looks up the description of the relation, "FLUSHON," and uses this information in conjunction with the location information provided by the FIND module to compute an "offset factor." The offset factor is then used to adjust the PICTURE module so that the image of the part, CUSHION, is generated at the correct size and location relative to the skeletal image.

The PUT module then returns to the object's propositional file and looks for another HASA entry, repeating the entire process. Depending on the task requirements, the PUT module may only add a few parts, or may generate as many parts as are stored. In the latter case, however, the parts initially generated may have faded away before the last part is added. That is, the image has a finite "capacity," defined by how quickly parts are generated and how quickly they fade; there comes a point when only so many parts can be present in the image at the same time.

The above description is only a brief overview of the workings of the actual algorithm, but should be sufficient to indicate just how complicated it is to mimic human performance on a computer. There are a number of reasons to go to the trouble of trying to spell out these details. First, one cannot be sure the theory can actually explain the results without doing so. Second, in the course of trying to specify the algorithm, alternative possible ways of proceeding become clear, pointing the way toward systematic new empirical research (instead of arbitrarily picking one alternative, one can try to discover how processing is actually done in the brain). Third, the algorithm allows one to make precise predictions about behavior in novel situations. These virtues will be illustrated in the following sections.

III. EXPLAINING THE RESULTS

The theory neatly explains the classes of empirical findings summarized above. The outlines of the explanations are as follows.

1. Effects of Number of Parts

Image generation times increase as increasing numbers of parts must be placed in the image. This result can be explained by assuming that parts are stored separately, and the PICTURE processing module is used to activate each part individually, with each operation requiring an additional increment of time.

2. Effects of Descriptions of Structure

Different ways of organizing a figure result in its image taking more or less time to generate, with time increasing when more parts are imposed on the stimulus. The conceptualization process presumably results in "top-down" parsing of the figure, so that different numbers of parts are stored. Again, each of these parts must be activated separately, resulting in an additional increment of time per additional part.

When the figure is a regular pattern, like an array, a single unit (e.g., a letter) is stored along with the description of how tokens of that letter are arranged. Thus, when later generating the image the description is used by the PUT processing module to arrange the letters correctly. Note, then, that image generation is hierarchical: the individual letters themselves can be composed using the PUT module (which arranges lines into a letter), which then arranges the individual letters into a pattern.

3. Effects of Descriptions of Relations

Images of scenes can be created when one hears or reads a description of the spatial relations among objects. The PUT processing module presumably uses the relations (after the words have been comprehended by language processing mechanisms), and uses them to calibrate the PICTURE processing module appropriately, as described above.

When people imagine letters of the alphabet in sequence, they prompt themselves by silently saying the letters. This result says nothing about the imagery process per se, but rather how we recall which items to image in sequence. This verbal prompt strategy may reflect the fact that many people memorize the alphabet using a rhyming chant, and hence the order is encoded in this form. The order of letters arranged into words is not memorized this way, so we do not use verbal prompts to recall them.

4. Effects of Size

Objects imaged at smaller sizes are usually formed more quickly than objects imaged at larger sizes. This result makes sense if the FIND processing module cannot locate the "foundation" part of small objects, and so fewer parts are actually generated into the image. This explanation leads us to expect that it should be *more* difficult to generate small images if one really tries to detail them completely, which is exactly what Farah and Kosslyn (1981) found.

5. Effects of Part Similarity

The more difficult it is to distinguish the parts, the more difficult it is to add one to the image. This result is explained if the FIND processing module must perform more exacting tests to identify a "foundation part" when parts are similar. For example, if one must only distinguish an *M* from a *G*, all one must look for is the presence of a straight or curved line; in contrast, if one is distinguishing an *M* from an *N*, now one must observe the number and arrangement of vertical and diagonal lines.

IV. PREDICTIONS

A variety of different predictions are made by the theory. In this section we will consider three of them, one based only on the claims about the nature of the processing modules; one based on the broad claims about the algorithm; and one based on details of the algorithm.

1. Predictions of Modular Decomposition

Although the Kosslyn and Shwartz theory is consistent with the data and the computational constraints, it is not the only theory that one could formulate. For example, we now assume that distinct functions are carried out by the PICTURE and PUT modules; however, these functions could in principle be carried out by a single module. Similarly, the functions carried out by the PICTURE and FIND processing modules could in principle be carried out by a single module. If the three processing modules posited by the theory are in fact distinct, we might be able to find that brain damage sometimes disrupts some of the modules while leaving others intact.

According to our theory, the PUT processing module is responsible for looking up and using descriptions of how parts are arranged together. This activity involves the manipulation of symbolic representations. It has long been claimed that the left cerebral hemisphere (at least in right-handed males) is used in symbol-manipulation activities (see Springer & Deutsch, 1981). Indeed, the right hemisphere appears to be deficient in the manipulation of symbolic representations. These notions led us to expect poor performance in the right hemisphere on tasks in which the PUT processing module must be used. In contrast, we had no reason to suspect that either hemisphere would be superior to the other in using the PICTURE or FIND processing module.

Thus, Kosslyn, Holtzman, Farah, and Gazzaniga (1985) asked whether there is a selective deficiency in the PUT processing module in the right cerebral hemisphere. To answer this question, they tested patients who had undergone "split brain" surgery for medical reasons; these people had their corpus callosa severed, leaving the two hemispheres physically separated and functionally isolated from one another (see Gazzaniga, 1970). If such a patient is asked to stare straight ahead, and a stimulus is flashed to the left side, the stimulus will be seen only by the right hemisphere (and only by the left hemisphere if it is flashed to the right side). Thus, by presenting stimuli, such as names of animals, off to one side, the investigators could test each hemisphere's ability to perform imagery tasks. For example, in one experiment the subjects were asked to respond by pressing one button if the named animal was larger than a goat, and another if it was smaller. When animals are close in size (e.g., hog versus goat), this task requires imagery (see Kosslyn, 1980, Chapter 9). In this case, however, only the global shape of the animal is needed; no parts are required to perform the task, Thus, it is of interest that both hemispheres performed virtually perfectly. In contrast, in another experiment the same stimuli were presented, but now the split-brain patient had to decide if the animal's ears protruded above the top of its skull (like a horse's) or did not protrude above the top (like a sheep's). This task does require arranging two parts in juxtaposition, and thus it is interesting that whereas the left hemisphere was extremely good at making the judgment, the right was at chance. Results like these supported the claim that the PUT processing module is functionally distinct from the other two.

2. Predictions About the Sequential Operation of Modules

Another set of predictions derives from the claim that parts are imaged sequentially. That is, the theory posits that the PICTURE module is used iteratively, imaging an object's parts separately and one at a time. To test this prediction, Kosslyn, Cave, and Provost (1985) adapted a technique developed by Podgorny and Shepard (1978). They showed people upper case letters that were drawn by filling in squares of a 4 x 5 grid. They then removed the filled squares and asked subjects to imagine the letter by mentally filling in the appropriate squares. They then presented x marks in two of the squares and asked whether those squares would have been filled had the letter been present. In the experiment, the x marks appeared only 300 msec after the cue was presented, which is not enough time to finish generating the image before the probes appeared; thus, one component of the response time will be the time to finish generating the image.

In addition to collecting times from the blank-grid task, subjects also performed the task when the letters were physically present in the grids. That is, the subjects were asked to determine whether or not both x marks fell on the now-visible letter. These times reflect the encoding (of the x marks), comparison (of marks and letter), decision (cover/noncover), and response components of the imagery task. Thus, by subtracting these times from the corresponding ones in the imagery task, an objective estimate of image generation time was obtained. The estimate of the difference in time to generate simple and complex letters obtained from this technique was almost identical to that found when subjects were asked simply to press a button when the letter was imaged in a blank grid.

The trick of this experiment was that Kosslyn, Cave and Provost varied the locations of the x marks. That is, the marks were positioned so that the letter would have to be more or less completed in order to "see" the x marks. The logic was that if letters are imaged segment by segment, then the position of the x marks will be critical; furthermore, if this variable reflects the time only to image successive segments, then this variable should not affect times in the perceptual condition, when the squares are actually filled in. And this is exactly what was found: When subjects imaged the letters, the further a mark was from the beginning (using the order in which the letters were usually drawn), the more time was required; but when subjects actually viewed the letters, there was no effect of x mark position whatever. Control experiments demonstrated that this effect was not due to subjects' imagining the letters all of a piece and then scanning over the image.

3. Predictions About the Algorithm

Another prediction stems from the details of the way in which parts are purportedly placed on images. According to the theory, a list of facts is searched until a part name is found. But this same list purportedly is searched to find a part even when imagery is not used. If so, then the same factors that affect the time to find a part in a simple question-answering task also ought to affect the time to image the part. That is, if one has to scan relatively far down the list to find the entry that an object has a part, times ought to be longer in question-answering tasks and in image-generation tasks than when the entry is higher on the list and less scanning is required. Furthermore, the magnitudes of the differences in time to respond to different items ought to be the same for a given item in both tasks.

To test this prediction, animal-property pairs were constructed in which both the size of the part and the association strength between the part and animal were varied (see Chapter 7 of Kosslyn, 1980, for a list of items). If association strength reflects the ordering of parts in a list (with more highly associated ones being "higher" in the list), as has sometimes been claimed, then subjects should more quickly affirm that animals have highly associated parts (e.g., stripes for zebra) than parts not highly associated (e.g., knees for zebra). This was, in fact, true. Furthermore, if our theory is correct, then the differences in time for the two kinds of items should be the same in the non-imagery task and the imagery task. This was also the case.

The results were not exactly the same in both cases, however: For the imagery task, more time was taken to "see" smaller parts, regardless of the association strength, replicating previous findings (Kosslyn, 1975); no such effect of size was found in the non-imagery condition. The effects of size presumably reflect the ease of using the FIND processing module, which only would be used here in the imagery condition. In addition, the imagery condition, as predicted, required more time overall, presumably because a host of additional processes follow the discovery that a part is available to be imaged on the object.

V. NEW DIRECTIONS

Two sorts of new directions are being pursued in developing the theory. On the one hand, we are filling in gaps in the theory; on the other hand, we are challenging some of the basic assumptions of the theory itself.

1. Filling in Gaps

Images are not of two-dimensional, flat surfaces; rather, they are of objects in three dimensions. Pinker (1980) explored the three-dimensional nature of images, but no one to date has explored how three-dimensionality is generated into an image. Is there a systematic difference in the order in which parts are generated depending on their relative depth from the "viewer"? Nearer parts may be defined with respect to the viewer's vantage point, whereas locations of more distant portions may be defined with respect to these nearer portions. Thus, parts of images corresponding to nearer portions of objects may be used as foundation parts because we know where they "go" in the image, and to these foundation parts may be added parts corresponding to more distant portions. Alternatively, the location might have an effect due to the order in which parts are initially encoded. If so, closer parts may be imaged first for this reason. Or perhaps intermediate portions may be at a "comfortable" viewing distance, and thus may be more likely to be initially encoded in perception, resulting in their being more likely to be generated initially into an image.

To investigate the idea that there would be a near-far ordering in image generation, Roth and Kosslyn (1985) employed a variant of the methodology developed by Kosslyn, Cave and Provost (1985), described above. Roth and Kosslyn required subjects to memorize the location of colored tiles on the floor and ceiling of a pictured corridor. The patterns extended from the tile nearest the viewer to the corridor's terminus in the distance. Later, subjects had to decide whether a single probe dot presented in an "empty" corridor would have fallen on or off the pattern had it been present. The logic was that if the image of the corridor was generated sequentially according to the near-to-far gradient, then subjects should be faster to affirm that a dot fell on a nearer portion of the pattern than when it fell on a more distant tile. And as in Kosslyn, Cave and Provost (1985), the perceptual times were subtracted from the corresponding imagery times, yielding a good measure of time to generate the image. The results suggest that people may, indeed, generate images by first filling in the nearest portions and then the farther ones. In fact, times increased nearly linearly with distance from the viewer.

Other related issues have yet to be examined. In particular, there is the question of whether given types of parts or properties (e.g., color and texture) are included only when they are needed to perform a task, or whether they are included obligatorily. Similarly, what about movement? Is the representation of movement continuous or a sequence of discrete steps? When is movement preserved in an image, and when does a "still" occur instead? What is the relationship between drawing and image generation? Are parts produced in the same order in both cases? (Preliminary data suggest that they might be.) In a different vein, we can ask even more basic questions such as, what information is included in a skeletal image? What angle is an object imaged at? How does one select a given version of an object to image?

2. Questioning Basic Assumptions

Although the theory has done reasonably well in explaining the "new" available results and in making new predictions, it clearly is not quite correct. Consider a very simple result: In one experiment, subjects are asked to categorize whether upper case letters have only straight lines or include any curved lines. Subjects are cued by seeing the lower case versions of the letters. For the first half of the experiment, only half of the letters of the alphabet were used; for the second half, the remaining letters were used. The simple finding is that times decrease during the first half as the letters are repeated, reflecting the effects of practice. The effects of practice during the first set of trials are evident even when we change the task halfway through, asking subjects to decide whether the upper case versions have a vertical line on the left; thus, these effects are not due to subjects' memorizing the responses (the new judgment required responses that were essentially uncorrelated with the earlier ones). But when the new letters were substituted halfway through the experiment, the effects of practice were eliminated and response times reverted to levels near those found at the outset of the experiment.

This finding is so intuitively plausible that it was a surprise when we realized that it had ominous implications for our theory. If practice effects are due to one of the three processing modules getting "warmed up" or the like, the effects should have transferred to the new letters. Instead, practice seemed "representation specific," which does not follow in a simple way from our theory.

3. An Alternative Architecture

One way of accounting for the available results preserves the decomposition into three processing modules, but modifies the nature of the algorithm at the most detailed level. Instead of assuming that images are stored as files of information in long-term memory, as in a computer, we can adopt an alternative functional architecture in which images are stored as patterns of weights on links in a network. That is, in this case images are formed by actually activating the representation, with the representation (pattern of weights) producing a distinctive pattern of activity in a neural network. (The weights themselves are determined by learning.) Because different images would be represented by different patterns, there is no reason to expect that "priming" of one pattern by repeated use would generalize to the priming of another (unless links happened to be shared by two representations). The details of such a model have not been worked out, but this kind of architecture is an interesting departure from that embodied in the imagery model and well worth exploring in depth.

Alternatively, perhaps a parallel architecture is used, but not a distributed parallel system. That is, perhaps the representations are nodes in a network, with each node representing a part or skeletal form. In this architecture, the nodes are hooked together by labeled links, with the labels indicating the relations among the parts (e.g., see Anderson, 1983). In such an architecture, it is convenient to assume that "spreading activation" is used to activate nodes. In this case, the PICTURE processing module would be taken as a description of how signals from nodes representing, perhaps, adjacent objects or properties cause other nodes to be activated, forming an image of the represented part or form. Because the relations between the objects are represented by labeled links, the PUT module would have a very different role than that posited earlier; the description of the module would now be taken as a description of how links "modulate" activation as it spreads to new nodes. That is, the process of activating the node could be affected so that the image is formed at different sizes and locations.

As the foregoing speculations should demonstrate, the theory of the architecture is very important because it introduces fundamental assumptions about how the system works. Although we have not examined these alternatives in detail, it seems clearly important to do so.

4. Two Types of Image Generation

Another issue that should be explored concerns the possibility that there is more than one way to generate images. In this case, the way one would generate images would depend on the nature of the object to be imaged. The theory we have described above makes good sense for objects that can undergo *nonlinear* transformations, but does not necessarily make good sense for objects that undergo only *linear* transformations. A linear transformation is one that operates uniformly on all parts of an object and is additive (i.e., two small transformations have the same result as one larger one). A cube is like this. Any given view or instance of a cube can be converted to any other view or instance by a combination of linear transformations (i.e., those that translate the object across the field, change its size, and/or change its orientation). Such objects tend to be rigid. For other objects, however, such linear transformations cannot be used to convert one version of the object to another. For example, a human figure can have its arms and legs in innumerable positions, with each part varying independently of the others. Similarly, the letter "*A*" can be drawn in one of many different ways, which cannot then be converted to other A's using linear transformations (e.g., the top can be square, rounded, or pointed; the letter can lean to the left, to the right, etc.). For these sorts of objects, which tend to be non-rigid, only the topographic relations among parts - what is connected to what, what is inside what - remain constant. Thus, it makes sense to represent the parts and relations of such objects separately, and to use nonpictorial, abstract relations to describe the parts' arrangement - just as we have done in our theory.

Our theory of image generation is very reasonable for objects that can undergo nonlinear transformations. Indeed, when most people are asked to imagine an upper case "A," they report that they are not imaging a specific letter they had seen previously (e.g., on the front page of yesterday's *Times*). Instead, they claim to be imaging a "generic" version of the letter. This introspection is exactly as expected if the image is being formed from a description of the relations among parts, which will result in a canonical version of the letter, a kind of "least common denominator," abstracted from the variability of individual instances. However, what about objects that undergo linear transformations and are rigid, with the parts never altering? Most aspects of specific faces are like this; facial expressions change very little, the parts never changing position. For objects like these, people do seem to image specific instances; a face, for example, will be imaged as it appeared at a specific time and place. To generate images of such objects, it may not be necessary to use propositional relations to coordinate parts into a single, composite image. Instead, parts may be encoded in absolute locations in space, with parts fitting on a global image much as different pictures can be placed on absolute locations on a table top (not relative to each other, only relative to the table top).

One way of testing the proposed distinction in processing between the two types of stimuli would be to discover whether the right hemisphere of split-brain patients, which has impaired symbol-processing ability, can generate images of multi-part, rigid objects (a capacity yet to be demonstrated), but not multi-part, nonrigid objects (as has been demonstrated in the past).

VI. CONCLUSIONS

The study of mental imagery has progressed very far since the days when simple introspection was the main methodology and common sense was the main theoretical framework. The present chapter has attempted to illustrate how computational constraints and computational theoretical concepts can be used in conjunction with empirical results. That is, we have tried to show how a logical analysis of the nature of the problem and what must be accomplished can aid in the development of a theoretical framework. One reason this is worth doing, we maintain, is that it leads one to ask questions in a systematic way, which in turn leads to fruitful empirical research.

Although we have offered a particular theory of image generation, we do not take these ideas as in any way finalized. As was developed in the preceding section of the chapter, there are numerous alternative approaches, and only empirical investigation will allow us to sort among them. The present theory, however, is probably a good approximation to a correct theory at a relatively high level of analysis. That is, it may well turn out that image generation can be described in terms of the three processing modules we have posited. Exactly how and when those modules need to be invoked, remain open questions. The attempt to answer these and other questions raised in this chapter mark the directions in which future research on image formation from a computational perspective is likely to proceed.

REFERENCES

Anderson, J.R. (1983). *The architecture of cognition*. Cambridge, MA: Harvard University Press.

Beech, J.R., & Allport, D.A. (1978). Visualization of compound scenes. *Perception*, 129-138.

Brooks, L. (1967). The suppression of visualization by reading. *Quarterly Journal of Experimental Psychology, 19*, 289-299.

Farah, M.J., and Kosslyn, S.M. (1981). Structure and strategy in image generation. *Cognitive Science, 4*, 371-383.

Finke, R.A. (1980). Levels of equivalence in imagery and perception. *Psychological Review, 86*, 113-132.

Finke, R.A., & Shepard, R.N. (in press). Visual functions of mental imagery. In L. Kaufman & J. Thomas (Eds.), *Handbook of perception and human performance*. New York: John Wiley and Sons.

Gazzaniga, M.S. (1970). *The bisected brain*. New York: Appleton-Century-Crofts.

Kosslyn, S.M. (1975). Information representation in visual images. *Cognitive Psychology, 7*, 341-370.

Kosslyn, S.M. (1980). *Image and mind*. Cambridge, MA: Harvard University Press.

Kosslyn, S.M. (1983). *Ghosts in the mind's machine*. New York: W.W. Norton.

Kosslyn, S.M., Cave, C.B., & Provost, D.A. (1985). Sequential processes in image generation: An objective measure. *ONR Technical Report #6*, Department of Psychology and Social Relations, Harvard University.

Kosslyn, S.M., Holtzman, J.D., Farah, M.J., & Gazzaniga, M.S. (1985). A computational analysis of mental image generation: Evidence from functional dissociations in split-brain patients. *Journal of Experimental Psychology: General, 114*, 311-341.

Kosslyn, S.M., Reiser, B.J., Farah, M.J., & Fliegel, S.L. (1983). Generating visual images: Units and relations. *Journal of Experimental Psychology: General, 112*, 278-303.

Kosslyn, S.M. & Shwartz, S.P. (1977). Simulating visual imagery. *Cognitive Science, 1*, 265-295.

Kosslyn, S.M., & Shwartz, S.P. (1978). Visual images as memory structures. In E.M. Riseman & A.R. Hanson (Eds.), *Computer vision systems*. New York: Academic Press.

Marr, D. (1982). *Vision*. San Francisco, CA: W.H. Freeman.

McGlynn, F.D., Hofius, D., & Watulak, G. (1974). Further evaluation of imagery latency and reported clarity as functions of image complexity. *Perceptual and Motor Skills, 38*, 559-565.

Moore, T.V. (1915). The temporal relations of meaning and imagery. *Psychological Review, 22*, 177-215.

Paivio, A. (1975). Imagery and syncronic thinking. *Canadian Psychological Review, 16*, 147-163.

Pear, J.J., & Cohen, R.G. (1971). Simple and complex imagery in individual subjects. *Psychological Record, 21*, 25-33.

Pinker, S. (1980). Mental imagery and the third dimension. *Journal of Experimental Psychology: General, 109*, 354-371.

Podgorny, P., & Shepard, R.N. (1978). Functional representations common to visual perception and imagination. *Journal of Experimental Psychology: Human Perception and Performance, 4*, 21-35.

Roth, J.D., & Kosslyn, S.M. (1985). Internal construction of the third dimension. *ONR Technical Report #8*, Department of Psychology and Social Relations, Harvard University.

Segal, S.J. (1971). Processing of the stimulus in imagery and perception. In S.J. Segal (Ed.), *Imagery: Current cognitive approaches*. New York: Academic Press.

Springer, S.P., & Deutsch, G. (1981). *Left brain, right brain*. San Francisco, CA: W.H. Freeman.

Weber, R.J., Kelley, J., & Little, S. (1972). Is visual imagery sequencing under verbal control? *Journal of Experimental Psychology, 96,* 354-362.

CHAPTER **6**

On Imagined Revolutions

Michael C. Corballis

I. INTRODUCTION

A revolutionary event in the experimental study of imagery took place just before 6:00 a.m. on November 16, 1968. Roger N. Shepard was in a state of hypnopompic suspension between sleep and waking at that time, and experienced a spontaneous image of three-dimensional structures turning in space (Shepard, 1978). This vivid experience led at once to the design of the first chronometric study of the mental rotation of three-dimensional objects. The experiment was subsequently carried out and reported in *Science* by Shepard and Metzler (1971).

In this experiment subjects were shown pictures of pairs of three-dimensional objects that were either the same, or else were enantiomorphs (mirror images). The objects were depicted in varying angular orientations, so that one object would have to be rotated in order to be matched with its "same" partner. In one condition this rotation was in the plane of the picture, while in another it was in the third dimension; examples are shown in Figure 1. The mean reaction time to judge that same pairs were, in fact, the same increased linearly with the angular orientation between the two, regardless of whether the orientations differed in the plane of picture or in the third dimension. This is shown in Figure 2. Shepard and Metzler took this to mean that subjects "mentally rotated" some internal representation of one of the shapes in order to match it against the other.

To many, this must seem an intuitively obvious result, and no major cause for alarm or excitement. However, to cognitive psychologists

151

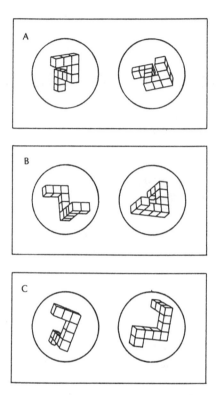

Figure 1. Examples of the shapes presented by Shepard and Metzler (1971). In (a), the shapes differ by a rotation in the plane of the picture. In (b), they differ by a rotation in the plane orthogonal to that of the picture. In (c), the two shapes are enantiomorphs. The subject must decide whether or not each pair depicts the same three dimensional object.

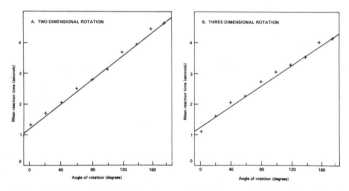

Figure 2. Reaction time plotted against the angular difference between the pairs of shapes, in the experiment by Shepard and Metzler (1971).

it was a shock, and sent waves through the discipline that have not yet subsided. In retrospect, perhaps we should be just as shocked that it did so, since mental rotation was by no means new to psychology. It has long been a critical component in the measurement of spatial ability (e.g., Thurstone, 1938). It also features in the Piagetian model of child development (Piaget & Inhelder, 1956, 1971) and, indeed, children are said to be unable to perform mental transformations of space until they reach the "concrete operational" stage of development at around 7 or 8 years of age.

Shepard's innovation was to use reaction time to measure mental rotation. This revealed a property of mental rotation that went against the grain of cognitive theory, although again it did not offend intuition. The linear function shown in Figure 2 suggests that mental rotation is analogous to *physical* rotation, in that both take longer the larger the angle of rotation. Although the experiment lacks the fine grain to prove it definitively, mental rotation seems to be smooth, in the sense that the representation passes continuously through all orientations between the initial and final ones. Shepard (1978) contrasts this with a computational process of rotation in which the intermediate steps would be psychologically meaningless. For example, a pattern might be represented as a set of coordinates that could then be rotated by a matrix multiplication, but if that multiplication were stopped at some intermediate stage, the representation would not in general have any coherent interpretation. Moreover, one would not expect matrix multiplication to yield the function shown in Figure 2.

Cognitive theory has been heavily influenced by the digital computer, and even those not working directly in artificial intelligence have for the most part assumed that cognition is fundamentally computational. Hence, the shock of Shepard and Metzler's elegant experiment. But, needless to say, the computational theorists have rallied. It is, of course, possible to model analogue processes on a digital computer to any desired level of precision, and Kosslyn and Shwartz (1977) have provided an explicit computational rendering of mental rotation that preserves the properties observed by Shepard and his colleagues.

Yet, the digital implementation of an analogue process of mental rotation seems cumbersome, and no programmer set the task of computing the new coordinates of a rotated pattern would solve the problem that way. Why, then, have theorists been so insistent on preserving computational models when an analogue model seems so much simpler? One reason, perhaps, is that it is difficult to envisage the brain as being other than computational, and so a computational answer is sought even if the data suggest that noncomputational alternatives might be simpler. The wheel is an efficient means of locomotion, yet it has never emerged in biological evolution; presumably, it is incompatible with the structure of living tissue. Similarly, there is nothing resembling a wheel inside the head, although such a device would provide an admirable solution to the problem of smooth rotation!

There are, of course, reasons for insisting that the brain does not model the world in analogue fashion. As Pylyshyn (1973) has argued, to do so would simply require too much storage and processing capacity, as the world is virtually infinitely variable in its manifestations. In modeling it, therefore, the brain must surely seek simplifying generalizations and rules, better conceptualized as propositions or descriptions than as pictures. If images are to be stored in memory, moreover, there seems no way to solve the problem of efficient retrieval unless the images are given some meaning or interpretation by which to file them. This again implies that images must be processed beyond the merely pictorial level.

Still, there are some respects in which a close, one-to-one modeling of the world seems both possible and desirable, especially with regard to those aspects of imagery most directly related to perception. Indeed, the distinction between perception and imagery is not sharp, and it is reasonable to suppose that they share common mechanisms (Hebb, 1968; Podgorny & Shepard, 1978). Mental rotation, for instance, may well involve processes that also occur during the perception of rotation. Now, it makes sense to suppose that during the perception of a rotating object, the perceiver maintains a continuous percept of the object as it rotates, just as one continuously monitors an object that moves about in space by making pursuit eye movements. If imagining is indeed derivative of perceptual activity, as Neisser (1976) has claimed, then it follows reasonably that imagining should retain the continuity of monitoring that is characteristic of perceiving.

It is useful at this point to distinguish between mental *processes* and mental *representations* involved in imagery. As a process, mental rotation may well be analogous to physical rotation in that it is smooth and time-bound. It need not follow, however, that the mental representation of an object or pattern that is mentally rotated is in any sense an analogue of the physical object or pattern it represents. Anderson (1978) has pointed out, for instance, that a pattern might be reduced to a set of descriptions in propositional format, in which there is but a single parameter representing the angular orientation of the pattern. Mental rotation might then be accomplished by incrementing or decrementing that parameter. Such a process preserves the smooth, analogue quality of rotation itself, while representing the pattern that is rotated as a set of propositions.

It seems unlikely, then, that there will be a simple, unique answer to the question of whether imagery is analogue or propositional in character. One might expect the long-term storage of images to be propositional, perhaps, while ongoing imaginal processes closely linked to perception might well retain an analogue quality. Even this generalization is likely to be overly simple, however. For instance, mental rotation may possess the smooth, time-bound quality of true rotation, yet there is no indication that it reflects other properties of true rotation, such as inertia. A representation might include both analogue and propositional elements, as a map displays spatial relations in analogue fashion yet also includes names or symbols that are, in essence, propositional (Anderson, 1978). To complicate matters further, Anderson argues that the issue of whether representations are analogue or propositional cannot be decided on the basis of behavioral evidence; any behavioral result can be modeled either in propositional or in analogue terms.

Should we, therefore, despair of making progress in experimental psychology of imagery? Elsewhere in this volume, Yuille argues that we should, indeed, abandon the experimental approach and seek solutions in more naturalistic settings. I agree that we should look to the ecological validity of our discipline, but I am not ready to abandon experiment. I shall argue, in fact, that experiments in mental rotation have clearly advanced our knowledge about imagery, even if they have not resolved the fundamental issue as to whether imagery is analogical or propositional in nature. Anderson also, in denying that this issue is resolvable in behavioral terms, has nevertheless urged that experimental work continue. He notes that physicists have been unable to resolve the issue of whether light consists fundamentally of waves or particles, but in tackling the question they have nevertheless learned a good deal about light.

The following selective review of experiments on mental rotation is organized around two broad issues. The first concerns the relations between imagining and perceiving. The second has to do with the special role of mental rotation in the discrimination of mirror images.

II. IMAGERY AND PERCEPTION

1. The Cooper-Shepard Paradigm

In a now classic series of experiments, Cooper and Shepard (1973) did much to clarify the interface between perception and imagery. They devised a mental-rotation task somewhat different from that of Shepard and Metzler (1971). Subjects were shown alphanumeric characters - specifically, the letters G, J, and R and the numerals 2, 5, and 7 - in varying angular orientations, and were timed as they decided whether each was normal or backward (i.e., mirror-reversed). If the subjects did not know in advance what each character would be or what orientation it would appear in, their reaction times increased sharply with the angular departure of the character from its normal upright orientation. This peaked orientation function is shown in Figure 3. Cooper and Shepard took it to mean that the subjects mentally rotated some internal representation of each character to the upright, and then compared it with an internally generated representation of that character in its normal (i.e., forward) version. Most subjects report that, introspectively, they do, indeed, mentally rotate the characters when performing this task.

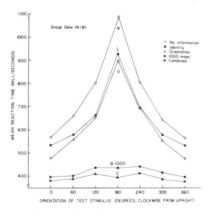

Figure 3. Reaction time to decide whether alphanumeric characters were normal or backward in experiment by Cooper and Shepard (1973). The different plots show the results for different kinds of advance information: with no advance information (N), information about identity only (I), or information about orientation only (O), the data show the characteristic mental-rotation function. With advance information about both identity and orientation (B-1000 and C), mental rotation was evidently no longer required.

At this point there are two reservations that might be noted. First, the peaked function shown in Figure 3 is not strictly linear. It tends to flatten out close to the 0° orientation, which gives it a slight downward concavity. This contrasts with the strict linearity observed in Figure 2. A likely, if somewhat *post hoc*, explanation for this flattening effect is that subjects do not always mentally rotate the characters the full angular distance to the upright, presumably because our internal representations or matching procedures can tolerate small angular discrepancies from the upright (Hock & Tromley, 1978). The nonlinearity of the function does not seem to threaten seriously the interpretation in terms of mental rotation.

The second reservation is that the *rate* of mental rotation implied by the function in Figure 3 is considerably higher than that implied by the functions in Figure 2, by a factor of about six or seven. In the task devised by Shepard and Metzler (1971) the estimated rate is some 50°-60° per s, while in that devised by Cooper and Shepard (1973) it is typically in the range of 350°-450° per s. Experiments described by Yuille (this volume) suggest that part of this variation is due to the fact that, in the first case, the representation to be rotated must be formed from a perceived stimulus, while in the second case the representations (of familiar alphanumeric characters) are already established in memory (see also Steiger & Yuille, 1983). Yuille argues that this and other sources of variation are sufficient to threaten the very viability of the concept of mental rotation.

This issue has also been discussed by Shepard and Cooper (1983). They refer to the work of Just and Carpenter (1976) and Carpenter and Just (1978) on the recording of eye-movement patterns in the comparison task that Shepard and Metzler (1971) used. Subjects frequently switched their gaze within and between stimuli while making the comparison. By studying the pattern and timing of these switches, Carpenter and Just were able to generate a model for how the task is broken down into component stages. This model implies a transformation-and-comparison process operating in discrete steps of about 50°. As Shepard and Cooper (1983) note "This view departs from our notion of mental rotation as an analogue process operating on holistic internal representations of visual objects" (p. 175). They go on to suggest that "experiments in which individually presented, well-learned stimuli must be compared with a long-term memory representation may encourage analogue processing of integrated internal representations, whereas experiments involving simultaneous comparison of unfamiliar objects may sometimes encourage more discrete, feature based rotational processing" (p. 175). Thus, ironically, the revolution begun by Shepard and Metzler (1971) may have been based on a false demonstration, and the paradigm developed by Cooper and Shepard (1973) may provide the truer example of an analogue mental transformation.

Let us now turn, then, to some of the properties of mental rotation in the Cooper-Shepard paradigm and, in particular, to the way in which they reveal the nature of the interactions between perception and imagery.

2. The Role of Advance Information

If subjects performing the Cooper-Shepard task are informed as to the identity of the character in advance of its presentation, but are not told its orientation, the reaction-time function is lowered overall but retains its peaked shape. A similar result occurs if the subjects are told the orientation of the character in advance, but are not told its identity. These results are shown in the graphs labeled "I" and "O," respectively, in Figure 3. Knowing the identity or the orientation (but not both) of the character in advance therefore does not eliminate the need to mentally rotate in deciding whether the character is normal or backward. This second result is of interest because it implies that people cannot mentally rotate an abstract frame of reference in anticipation of some unknown stimulus in a known orientation (Cooper & Shepard, 1973).

There is evidence, however, that the orientational frame of reference can be adjusted for head-tilt in advance of stimulus presentation. My colleagues and I have studied mental rotation using the Cooper-Shepard task when subjects tilted their heads 66° to the left or right (a head-tilt of 66° leaves the eyes tilted at roughly 60°). The functions were essentially unaltered by head-tilt, implying that subjects were able to adopt a gravitational rather than a retinal reference frame (Corballis, Zbrodoff, & Roldan, 1976). In this experiment, the characters were displayed on a screen set in a room that was slightly darkened, but clearly visible. Under more restricted conditions, with environmental cues removed, the mental-rotation function may be influenced by head-tilt. In a further study of the effects of head-tilt under such conditions, we have found the subjective vertical in mental rotation to lie between the gravitational and retinal verticals, but to vary somewhat depending on the physical surround (e.g., whether it is circular or square) and on the degree of head-tilt (Corballis, Nagourney, Shetzer, & Stefanatos, 1978). We concluded that the reference frame in mental rotation need be neither gravitational nor retinal, and that it is located by automatic rather than voluntary processes.

Although subjects seem to have little or no voluntary control over the abstract frame of reference, they are evidently able to prepare an image of a particular, known stimulus in advance of presentation. When Cooper and Shepard (1973) informed subjects as to *both* the identity *and* the orientation of the character to be presented, the reaction-time function flattened out, as shown in the graphs labeled "C" and "B-1000" in Figure 3. For this flattening to occur, it was necessary to present the advance information some 1 s in advance of the stimulus, presumably in order to give the subject time to generate an image of the character and mentally rotate it to the required orientation.

This result shows that an alphanumeric character imagined in a particular orientation can serve in lieu of the character actually perceived in that orientation, at least for the purposes of matching - a finding that gives empirical support to Neisser's (1976) conception of an image as an anticipated percept. This is even more strikingly illustrated by another experiment carried out by Cooper and Shepard (1973) in which they instructed subjects to imagine a character in successive orientations. The verbal prompts "up," "tip," "tip," "down," "tip," "tip," were spoken at half-second intervals in synchrony with the appearance of small tick marks at six locations around the border of a circular field in 60° steps, starting at the top. These signals served to notify the subject as to the required orientations of the image at each half-second interval. The character was then actually presented at a point in time coincident with one of these prompts, and the task was to decide whether it was normal or backward. Reaction time increased, not with the angular distance of the character from the normal upright, but with the angular distance of the presented character from the imagined one. This again suggests that percept and image occupy the same representational space.

Although this last result seems also to suggest that mental rotation is smooth, Cooper and Shepard conceded that their subjects may have rotated their images in discrete 60° steps from one orientation to the next. However, Cooper (1976) later carried out a refinement of the experiment in which the test stimulus could appear at orientations intermediate to the prompted ones. The results suggested that the image retained its integrity at these intermediate orientations. Experimental evidence is thus converging on the view that mental rotation does have the continuous property of physical rotation.

3. Perceived and Imagined Rotation

The experiments just reviewed show that a stimulus imagined in a particular *orientation* has at least some of the properties of the same shape perceived in that orientation. But does mental rotation itself have any of the properties of perceived *rotation*? That is, is there any dynamic interaction between perceived and imagined rotation? Do the two processes share any common neural substrate? We have tried to address this issue by studying the influence of a perceptual after-effect of rotation upon mental rotation.

Our technique is to have subjects watch a rotating textured disk in between presentations of the alphanumeric characters to be judged normal or backward. This induces a motion after-effect such that each character, when presented, is perceived as rotating in the direction opposite to that of the disk. It is important to note that the motion after-effect has the paradoxical property of causing the stimulus to move without changing its position (e.g., Rock, 1975; Wohlgemuth, 1911); thus, the rotating disk causes the presented character to rotate without changing its perceived orientation! Any influence on mental rotation should therefore be due to the interaction between perceived and imagined *rotation* and not to any perceived change in orientation.

Corballis and McLaren (1982) showed that the reaction-time function in the mental-rotation task was, indeed, skewed in opposite directions by opposite motion after-effects. In particular, the maxima were shifted one step along the orientation axis. Thus, with a clockwise after-effect, the maximum reaction time occurred not when the characters were oriented at 180°, but when they were oriented at 120°, and with an anticlockwise after-effect the maximum reaction time was to characters oriented at 240°. There was relatively little influence on the minima. As expected, the data were consistent with the interpretation that the influence was on the rotation component itself, and not on the perceived orientation of the characters.

For her Masters thesis at the University of Auckland, McLaren (1983) carried out a more exacting study of the influence of a rotation after-effect on mental rotation. Unlike Corballis and McLaren (1982), she included a control condition in which the task was stationary. She reduced the interval of disk inspection between stimulus presentations from 9.5s to 6.5 s, but increased the number of presentations per condition for each subject from 72 to 144. The disk was rotated at 25 rpm, and each subject received the 144 trials under each rotation condition (clockwise, anticlockwise and stationary) in counterbalanced order. There were nine subjects, and each stimulus was presented for 2 s. McLaren's results are plotted in Figure 4.

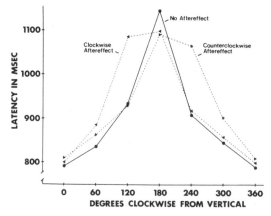

Figure 4. Mental rotation functions under the influence of a clockwise, anticlockwise, or no aftereffect, in the experiment by McLaren (1983).

It is clear that the after-effect did influence the reaction-time function. This influence was not one of translation along the horizontal axis, however, which means that the after-effect did not simply alter the perceived orientation of the characters. Moreover, the after-effect did not simply combine in additive fashion with mental rotation, since reaction time was not reduced when the after-effect was in the same direction as the presumed direction of mental rotation. In fact, the only statistically significant effect of the after-effect was to increase reaction time when it was in the direction opposite to that of the presumed mental rotation, and then only when the stimuli were oriented at 120° from the upright.

One possible explanation of these results is that subjects sometimes mentally rotated the characters through the larger angle to the upright when the after-effect was in that direction. In particular, when the letters were oriented at 120° from the upright, the subjects may have sometimes rotated them through 240° - the long way round - if this was the direction of the after-effect. In a follow-up study, McLaren asked the subects to report the direction of mental rotation after each trial. The results did not support the "larger angle" theory, although they were not entirely definitive since the requirement to report the direction reduced the influence of the after-effect. Moreover, there may be some question as to whether subjects genuinely have introspective access to the direction of mental rotation; rather, the subjective impression of mental rotation may be a rationalization rather than a true experience of a mental process.

But whatever the mechanism, perceived and imagined rotation clearly do interact. This further substantiates the idea that perception and imagery share a common representational space, and lends further credence to the notion that there is a truly rotational element underlying mental rotation.

III. MENTAL ROTATION AND MIRROR IMAGES

Nearly all studies of mental rotation have involved the discrimination of mirror images. In the paradigm used by Shepard and Metzler (1971), the patterns to be compared were either the same or else they were mirror images of one another. In the Cooper-Shepard task, the subject is required to discriminate whether a given shape is the same or the mirror image of some standard or canonical form. Mental rotation is typically *not* induced if the task is simply to identify or categorize a disoriented shape, regardless of its left-right orientation (Corballis & Nagourney, 1978; Corballis, Zbrodoff, Shetzer, & Butler, 1978; Eley, 1982, 1983; White, 1980).

The discrimination of mirror images requires the ability to tell left from right and this, in turn, requires reference to the left and right sides of our bodies. If there were no distinction between our left and right sides - that is, if we were perfectly bilaterally symmetrical - we would be unable to perform the Cooper-Shepard task. To see that this is so, imagine a bilaterally symmetrical person trying to decide whether letters in varying angular orientations are normal or backward. Now suppose the whole situation is mirror-reversed, as if viewed through a looking glass. The forward letters would then become backward ones and the backward ones forward ones, but the *person,* being bilaterally symmetrical, would remain quite unaltered. Clearly, the task would be quite beyond the powers of such a person, who has no way of telling whether the world is reversed or not (see also Corballis & Beale, 1970, 1976, 1983).

Another way to state the problem is to note that the distinction between enantiomorphs, or between left and right, cannot be expressed in purely symbolic form. There is no way to send a coded message to a recipient on some distant planet, say, with instructions on how to make a right-handed glove, and be sure that the product will not be a left-handed glove. This is what Martin Gardner (1964) has called the "Ozma problem": How to communicate symbolically our meaning of the terms "left" and "right." William James (1890) long ago recognized that left and right could only be distinguished in an analogue, nonsymbolic fashion:

If we take a cube and label one side *top*, another *bottom*, a third *front*, and a fourth *back*, therer remains no form of word by which we can describe to another person which of the remaining sides is *right* and which *left*. We can only point and say *here* is right and *there* is left, just as we should say *this* is red and *that* blue (p. 150; his italics).

If the discrimination of mirror images requires reference to our own left-right asymmetry, the role of mental rotation becomes understandable. It serves to align the shape with our own bodily coordinates, so that its left-right asymmetry, or parity, may be discerned. But, note that this implies that mental rotation serves the same function as *physical* rotation. This argument thus reinforces Shepard's (1978) view that mental rotation has an analogue quality; at some level of representation, a shape that is imagined in an upright orientation is the same as that of the shape actually perceived in that orientation.

As mentioned above, studies of mental rotation using the Cooper-Shepard paradigm have required subjects to discriminate mirror images. These shapes have included mirror-image alphanumeric characters (Cooper & Shepard, 1973), left and right hands (Cooper & Shepard, 1975), mirror-image polygons (Cooper, 1975), mirror-image letter-like symbols (Eley, 1982), and so on. Moreover, subjects performing these tasks have always treated the shapes as *left-right* mirror images. In discriminating a normal from a backward R, for instance, subjects invariably mentally rotate the backward R so that it is a left-right mirror image of a normal R and not, say, an up-down mirror image of it. However, even if the task were to involve the discimination of shapes whose canonical forms were up-down mirror images (or mirror images about *any* axis), it would still require reference to left and right.

This can be illustrated with reference to a study reported by Corballis and McLaren (1984). Subjects were required to make various timed discrimination among the lowercase letters b, d, p, and q, presented singly in varying angular orientations. Because of the nature of these letters, notoriously confusing to children first learning to read, the stimuli remained constant while the required discriminations changed. The reaction-time functions indicated that the subjects generally adopted a mental-rotation strategy whether the required discrimination was between letters whose canonical forms were left-right mirror images (b versus d; p versus q) or between letters whose canonical forms were up-down mirror images (p versus b; d versus q).

Consider the problem, for instance, of deciding whether a disoriented letter is "b" or "p." In order do this, one must understand how it would look if the rounded portion were to the *right*, and the straight portion to the *left*; only then can one decide whether the rounded portion is at the top (p) or bottom (b) of the straight portion. Hence, the task still requires reference to left and right, and the fact that the subjects mentally rotated the letters to their upright orientations suggests again that reference to left and right was achieved in analogue fashion.

In another study carried out with the help of Sharon Cullen, subjects were shown letters that were either vertically symmetrical (A, T, U, or V) or horizontally symmetrical (B, C, D, or E), in varying angular orientations. An asterisk was placed to the left, right, top, or bottom of each letter (relative to its own coordinates), and the subjects were timed as they called out "left," "right," "top," or "bottom," as appropriate. Figure 5 shows the reaction-time functions under the different conditions.

Decisions as to whether the asterisk was to the left or right evidently required mental rotation, as shown by the characteristically peaked functions in the right-hand panel of the figure - although there is rather more flattening of the function around the vertical than is evident from studies requiring discrimination of mirror-image shapes. Mental rotation was apparently required even when the discrimination was not strictly a mirror-image one, as in the case of the letters B, C, D, and E, where left and right sides could have been discriminated in terms of the distinctive features on either side of these letters rather than with reference to left and

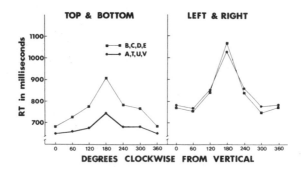

Figure 5. Reaction time as a function of angular orientation for judgments about the top and bottom (left panel) and about the left and right (right panel) of disoriented letters.

right. The use of the *labels* "left" and "right" seems to have been sufficient to induce mental rotation.

Decisions as to whether the asterisk was at the top or bottom of the letters only required mental rotation when the discrimination was a mirror-image one; that is, when the letters were horizontally symmetrical (B, C, D, or E). When the tops and bottoms were distinguished by critical features (A, T, U, or V), mental rotation did not occur. These results can be seen in the left-hand panel of Figure 5. They illustrate again the role of mental rotation in mirror-image discrimination even if the mirror images, in their canonical orientations, are not left-right ones.

In order to show that it is the *discrimination* of left and right, or of top and bottom, that is critical, we also carried out a control experiment in which subjects were required to make one response if the asterisk was to the left *or* right and another if it was on the top *or* bottom. As Figure 6 shows, there was no evidence of mental rotation at all with this task. It is only when the subjects must decide which is the left or right side, or in the

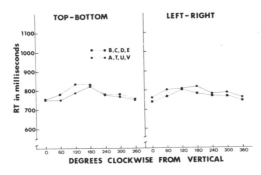

Figure 6. Reaction time as a function of angular orientation in an experiment requiring one response if an asterisk was on the top or bottom (left panel), and another response if it was to the left or right, of disoriented letters.

case of horizontally symmetrical letters, which is the top or bottom, that they resort to mental rotation.

Mental rotation can thus be regarded as our internal, analogue solution to the Ozma problem. It enables us to make a distinction that cannot be made in abstract, symbolic terms. This does not mean that images must be wholly analogue in form; like maps, they may have both analogue and propositional characteristics. The recognition that a given letter is an R, for instance, may depend purely on analysis of its features, codable in symbolic form, but the recognition that it is a *forward* R as distinct from a backward one has an inevitable analogue component.

IV. CONCLUSIONS

Shepard's hypnopompic vision represented a turning point, so to speak, in the development of cognitive psychology. There had, of course, already been a considerable amount of work on imagery, although much of it had been in the context of verbal behavior (e.g., Paivio, 1971). Even mental rotation was not a new phenomenon - as noted earlier, it had been studied by the Piagetians and the psychometricians. But, to cognitive psychology it represented a swing away from the emphasis on symbolic representation and computational processes that had so dominated the field.

By and large, research on mental rotation has continued to support Shepard's original claims about the nature of imagery, while at the same time expanding our empirical knowledge about imagery, perception, and the relations between them. As a *process*, imagining clearly has much in common with perceiving; indeed it seems reasonable to conclude that imagining *is* the cognitive component of perceiving, but is dissociable from the sensory component. Thus, a prepared image can serve in lieu of a stimulus and modify the processing of subsequent stimuli; conversely, the perception that a stimulus is rotating can influence subsequent mental rotation of that stimulus. Experiments on mental rotation are as much experiments on perception as experiments on imagery.

The special role of the left-right dimension in mental rotation also lends support in a subtle way to Shepard's claims about the analogue nature of the image. The left and right sides of an object or pattern can only be established with reference to one's own left and right sides. That this can be accomplished by *either* physical *or* mental rotation is thus further testimony to the close correspondence between the two. It may not be too far-fetched to suppose that the internal representation of a rotated stimulus in the Cooper-Shepard paradigm is mapped in analogue fashion across the brain, so that its left and right sides are referred physically to the left and right sides of the brain. There is some evidence that spatial images are mapped in this fashion from the work of Bisiach and Luzzatti (1978) on unilateral hemineglect following right-hemispheric brain injury. These authors found that patients tended to neglect the left sides of *imagined* scenes as well as perceived ones. In work carried out at the University of Auckland, Ogden (1983) has reported both left-and right-sided neglect of images among patients with unilateral lesions. We plan now to test whether patients suffering unilateral brain lesions might show neglect of one or the other side of an image *that has been mentally rotated.*

In this review I have taken a narrow approach, focusing on mental rotation itself, despite the fact that there are, of course, many other ways of studying the properties of imagery. In this I may seem to set myself deliberately in opposition to Yuille (this volume) in his quest for a broader, more ecological foundation for our science. Our experimental subjects sit in a darkened room, watching stimuli of a kind they are unlikely to encounter in their everyday lives flashing briefly before their eyes, and making staccato responses that are totally devoid of poetry, if not of emotion. What could be further removed from the real world?

However, I do not think that laboratory experiments in psychology are intended as microcosms of the real world, just as the test tube in chemistry or biology was not intended to encapsulate physical or biological reality in their natural manifestations. The very purpose of experiment is to remove phenomena from their natural settings and study them under controlled and usually unnatural conditions. This approach has proven highly effective in the physical and biological sciences, and has almost certainly had more profound "real-world" consequences than a more naturalistic, observational approach. Of course, these consequences have not always proven desirable; we need only think of nuclear fission, gene splicing, or biological warfare to be reminded of, at least, the potential for evil.

I have tried to show that research on mental rotation has progressed a good deal even since the now classic studies of Shepard and his colleagues, and that the issues it touches upon are surprisingly diverse. As a result of these studies, I think we now know more about perception as well as about imagery, and about the temporal and spatial structure of human thought. Experimental psychology offers no immediate solutions to such pressing problems as poverty or East-West detente, but then it seems also unlikely to contribute substantially to the Orwellian nightmare. In my case study of mental rotation, I have tried to show simply that we are capable of making progress, albeit slow, toward an understanding of the workings of the human mind.

REFERENCES

Anderson, J.R. (1978). Arguments concerning representations for mental imagery. *Psychological Review, 85*, 249-277.

Bisiach, E., & Luzzatti, C. (1978). Unilateral neglect of representational space. *Cortex, 6*, 129-133.

Carpenter, P.A., & Just, M.A. (1978). Eye fixations during mental rotation. In J.W. Senders, D.F. Fisher, & R.A. Monty (Eds.), *Eye movements and the higher psychological functions.* Hillsdale, NJ: Lawrence Erlbaum Associates.

Cooper, L.A. (1975). Mental rotation of random two-dimensional shapes. *Cognitive Psychology, 7*, 20-43.

Cooper, L.A. (1976). Demonstration of a mental analog of an external rotation. *Perception & Psychophysics, 19,* 296-302.

Cooper, L.A., & Shepard, R.N. (1973). Chronometric studies of the rotation of mental images. In W.G. Chase (Ed.), *Visual information processing.* New York: Academic Press.

Cooper, L.A., & Shepard, R.N. (1975). Mental transformations in the identification of left and right hands. *Journal of Experimental Psychology: Human Perception and Performance, 104,* 48-56.

Corballis, M.C., & Beale, I.L. (1970). Bilateral symmetry and behavior. *Psychological Review, 77,* 451-464.

Corballis, M.C., & Beale, I.L. (1976). *The psychology of left and right.* Hillsdale, NJ: Lawrence Erlbaum Associates.

Corballis, M.C., & Beale, I.L. (1983). *The ambivalent mind.* Chicago, IL: Nelson-Hall.

Corballis, M.C. & McLaren (1982). Interaction between perceived and imagined rotation. *Journal of Experimental Psychology: Human Perception & Performance, 8,* 215-224.

Corballis, M.C.,& McLaren, R. (1984). Winding one's ps and qs: Mental rotation and mirror-image discrimination. *Journal of Experimental Psychology: Human Perception & Performance, 10,* 317-318.

Corballis, M.C., & Nagourney, B.A. (1978). Latency to categorize disoriented alphanumeric characters as letters or digits. *Canadian Journal of Psychology, 23,* 186-188.

Corballis, M.C., Nagourney, B.A., Shetzer, L.I., & Stefanatos, G. (1978). Mental rotation under head tilt: Factors influencing the location of the subjective reference frame. *Perception & Psychophysics, 24,* 263-273.

Corballis, M.C., Zbrodoff, N.J., & Roldan, C.E. (1976). What's up in mental rotation? *Perception & Psychophysics, 19,* 525-530.

Corballis, M.C., & Zbrodoff, N.J., Shetzer, L.I., & Butler, P.B. (1978). Decisions about identity and orientation of rotated letters and digits. *Memory & Cognition, 6,* 98-107.

Eley, M.G. (1982). Identifying rotated letter-like symbols. *Memory & Cognition, 10,* 25-32.

Eley, M.G. (1983). Symbol complexity and symbol identification with rotated symbols. *Acta Psychologica, 53,* 27-35.

Gardner, M. (1964). *The ambidextrous universe.* New York: Basic Books.

Hebb, D.O. (1968). Concerning imagery. *Psychological Review, 75,* 466-477.

Hock, H.S., & Tromley, C.L. (1978). Mental rotation and perceptual uprightness. *Perception & Psychophysics, 24,* 529-533.

James, W. (1890). *Principles of psychology. Vol. II.* New York: Henry Holt & Co.

Just, M.A., & Carpenter, P.A. (1976). Eye fixations and cognitive processes. *Cognitive Psychology, 8,* 441-480.

Kosslyn, S.M. & Shwartz, S.P. (1977). A data-driven simulation of mental imagery. *Cognitive Science, 1,* 265-296.

McLaren, R. (1983). *Interactions between perceived and imagined rotation.* Unpublished master's thesis, University of Auckland, New Zealand.

Neisser, U. (1976). *Cognition and reality.* San Francisco: W.H. Freeman.

Ogden, J.A. (1983). *Out of mind, out of sight: Unilateral spatial disorders in brain-damaged patients.* Unpublished doctoral dissertation, University of Auckland, New Zealand.

Paivio, A. (1971). *Imagery and verbal processes.* New York: Holt, Rinehart, & Winston.

Piaget, J., & Inhelder, B. (1956). *The child's conception of space.* New York: Basic Books.

Piaget, J., & Inhelder, B. (1971). *Mental imagery in the child.* New York: Basic Books.

Podgorny, P., & Shepard, R.N. (1978). Functional representations common to visual perception and imagination. *Journal of Experimental Psychology: Human Perception & Performance, 4,* 21-35.

Pylyshyn, Z.W. (1973). What the mind's eye tells the mind's brain: A critique of mental imagery. *Psychological Bulletin, 80,* 1-24.

Rock, I. (1975). *An introduction to perception.* New York: Macmillan.

Shepard, R.N. (1978). The mental image. *American Psychologist, 33,* 125-137.

Shepard, R.N., & Cooper, L.A. (1983). *Mental images and their transformations.* Cambridge, MA: The MIT Press.

Shepard, R.N., & Metzler, J. (1971). Mental rotation of three-dimensional objects. *Science, 171,* 701-703.

Steiger, J.H., & Yuille, J.C. (1983). Long-term memory and mental rotation. *Canadian Journal of Psychology, 37,* 367-389.

Thurstone, L.L. (1938). *Primary mental abilities. Psychometric Monographs, 1,* Pp. 121.

White, M.J. (1980). Naming and categorization of tilted alphanumeric characters do not require mental rotation. *Bulletin of the Psychonomic Society, 15,* 153-156.

Wohlgemuth, A. (1911). On the aftereffect of seen movement [Monograph Supplement]. *British Journal of Psychology, 1, 1-117.*

CHAPTER 7

The Conceptual Basis of Cognitive Imagery Models: A Critique and a Theory

Geir Kaufmann

I. INTRODUCTION

A natural consequence of the redesign of psychological inquiry within the new cognitivist paradigm is that the subject of mental representation in cognition has moved into the center of the theoretical debate (see e.g., Block, 1982; Mehler, Walker & Garrett, 1982). Thus, a leading theoretical entrepreneur of the cognitive school, Jerry Fodor, argues that cognitive psychology is inevitably committed to the postulation of internal representational systems. Since cognitive processes must have a medium to work in, we have to attribute representational systems to whatever organisms are endowed with cognitive processes. A primary research goal of cognitive psychology is, thus, to elucidate "..... the intelligent management of internal representations" (Fodor, 1976, p. 164). Another of the frontrunners, Margaret Boden, has argued the same point (Boden, 1979).

Working from a neo-associationist theoretical basis, Allan Paivio voices the same general spirit when he defines a new branch of this tradition, which he coins "neomentalism." The task of neomentalism is to be "the objective study of the structure, function, and development of mental representations" (Paivio, 1975a, p. 264).

This may be seen as a fortunate turn of events for at least two reasons. Since human cognition is basically *symbolic* cognition, it seems to be a sound way of making up the research priorities. Secondly, when basically different general theoretical schemes are addressed to exactly the same issue, a clarification of the validity of their basic assumptions should be accelerated.

With such a prescription for a fruitful research program, imagery is certainly quite a tidbit on the theoretical menu. In fact, the current debate over the nature and functional properties of mental representations has, in large measure, taken as its point of departure the conceptual problems attaching to imagery as a symbolic system (e.g., Anderson, 1978; Anderson & Bower, 1973; Block, 1982; Chase & Clark, 1972; Hayes-Roth, 1979; Kosslyn, 1980, 1981; Paivio, 1975a; Pylyshyn, 1973, 1979, 1981).

In this connection, it is interesting to note that the philosopher Alistair Hannay (1971) has suggested that the conceptual problems relating to the nature of imagery may lay the ground for an *experimentum crucis* on the validity of different philosophical theories of mind. According to Hannay, a clarification of such concepts as "mental image," "picturing," "seeing in the mind's eye," etc. is a particularly strategic way of getting to the basic premises of different theories. We may hope that the study of imagery will similarly uncover the basic premises of the various psychological theories of the nature of mental representations and spur research that may lead to real progress in our understanding of these intricate matters.

In view of the recent proliferation of theoretical models in the field of imagery (e.g., Hampson & Morris, 1978; Kieras, 1978; Kosslyn, 1980, 1981; Marks, 1977, 1983; Neisser, 1976), it may seem to be the wrong maneuver to place still another candidate on the scene rather than attempting to settle the dispute between the existing alternatives on empirical grounds, as Kosslyn and Pomerantz (1977) advocate. Although all of the existing theories focus on important aspects of imagery, it seems to me, however, that none of the theories to be discussed in this chapter escapes serious logical incoherences.

The major aim of the present article, then, is to develop an alternative, logically coherent view on the conceptual status of imagery as a symbolic system, which is compatible with empirical evidence at hand. Heading for this goal, we will proceed along the following path: To get to the roots of the assumptions behind the main theories currently in circulation, we will trace their historical-philosophical origins. Hopefully, this will give us a firm and comprehensive basis for a critical evaluation of the core ideas of the theories in their present form. The conclusions drawn will form the point of departure for working out an alternative theoretical position. Although the main concern of the present inquiry is with the conceptual aspects of imagery theory, we will wind up our discussion by examining the validity of some general empirical implications which follow directly from the theory.

The theoretical territory covered in this chapter is occupied by three of the major groups of imagery models.

(1) *The analog theory.* In this theory, images are seen as analog representations of perceptual information. The image representation is held to correspond rather directly to concrete objects and situations, and to symbolize by way of resemblance to the object it "stands for." The conscious image is given a functional role in cognition by virtue of its quasi-perceptual properties (Kosslyn, 1980, 1981; Paivio, 1971, 1975b; Shepard, 1978a, 1978b).

(2) *The propositional theory.* Both verbal and nonverbal information is held to be represented in a common, amodal format. The real (or at least most important) cognitive work is seen to take place at this abstract and conceptual level of representation. Images are either regarded as epiphenomenal or assigned a secondary, supportive function in cognition (Anderson & Bower, 1973); Chase & Clark, 1972; Pylyshyn, 1973, 1981).

(3) *Anticipation theory.* Images are assigned a dispositional status and conceived as perceptual anticipations (Neisser, 1976).

II. ANALOG REPRESENTATIONS AND THE IMAGIST THEORY OF CLASSICAL EMPIRICISM

The analog theory has firm roots in the imagist theory of classical empiricism (see Price, 1969, for a comprehensive exposition). Since many of the basic conceptual ideas have been carried over into contemporary variants of the imagery, it may be worthwhile to recapitulate its main features, as well as the charges that have been brought against it.

In the imagist theory the mental image is given an epistemological key role. Its central thesis is that mental images, and especially *visual* images, constitute the very bedrock of cognition. The most clear-cut and systematically developed version of the theory was formulated by Berkeley (1710). In Berkeley's theoretical system, a distinction is drawn between *primary* and *secondary* symbolic systems. Mental images are the primary symbols, whereas all other symbols are secondary and derived. Among the secondary symbolic systems, language is the most important. Words get their meaning indirectly through their relation to images. Words, then, are *substitutes* for images. As substitutes, words may have an important function in cognition, since they can be manipulated more easily and quickly than can images. The theory does not entail, then, that thinking necessarily is done in images, as it is often interpreted (e.g., Horowitz, 1970). Rather to the contrary, Berkeley claims that thinking occurs mainly in the medium of words, but that thinking is only *meaningful* when the words are convertible, either directly or indirectly, into the relevant images.

The imagist conception of thinking was transferred intact to the structuralist psychology organized by Wundt and Titchener: Images form the basic substrate of thinking, whereas language is seen as a secondary symbolic system which it is expedient to employ as a substitute for images (e.g., Titchener, 1910, 1914).

Why such primacy to images? As pointed out by the philosopher Price (1969), the thesis follows as a logical consequence of the empiricist standpoint: Thinking must be related to reality. Since thinking is mainly cognition in absence of direct contact with the objects in reality, images have the clear advantage over words in being nearer to the objects and situations they represent. Images serve, then, as the fundamental link between thinking and reality, and so constitute the basic meaning substrate of language.

This line of reasoning invites a host of serious problems. The basic assumption is that we have a direct and immediate access to an independently given reality. The foundation for objective knowledge is, then, provided by way of the immediate sensations being represented in images in pure, undistorted form. Consequently, concept formation is regarded as a matter of abstracting recurrent features directly given in sense experience. But, as many philosophers have convincingly argued (e.g., Geach, 1971; Rorty, 1980; Wittgenstein, 1953), there is nothing unambiguously given in sense experience. The content of our sense experience may always be interpreted in different ways, and the meaning is constructively conferred *upon* it, rather than directly given *in* it. The point here becomes very clear when we consider the problems involved in the image theory of meaning. If a word elicits an image, it will always contain many features that are irrelevant to its meaning function. Let us say that the word "five" evokes the image of a hand with five fingers. We know that the form, color and spatial arrangements of the fingers do not matter as far as the meaning is concerned. But this means that *the image itself has to be interpreted*. The conclusion must be that the image cannot fulfill any basic meaning function, and we have only pushed the problem of meaning one step further away. Moreover, it can be shown that the grammar of "understanding the meaning" and "having images" is totally different (see, e.g., Baker & Hacker, 1980). A person's avowal concerning his images is authoritative. His avowal concerning his understanding may well be contested on the basis of his subsequent performance. Furthermore, while it is possible to keep an image in mind for two and one-half minutes (see McKellar, 1957), it does not make similar sense to talk of keeping an understanding in mind for two and one-half minutes. The conclusion must be that the meaning of a word cannot be equated with imagining its denotation.

In addition to the problems related to the semantic dimension of imagery, difficulties also accrue to the *conceptual* component of the image. How is it possible for images to function as *general* symbols, representing classes of objects and events? The problem here is twofold: The image resembles one thing too closely, and at the same time resembles too many things (an image of a dog may resemble, say, an Alsatian; at the same time it also resembles wolves, foxes, animals, organisms, etc.). Thus, the image seems to be both too *specific* and too *ambiguous* to function as a general, conceptual representation. The problem is a serious threat to the imagist theory. Since verbal symbols easily meet the requirements for conceptual representation, and at the same time are regarded as secondary and derived from images, a convincing account of the way an image can function as a general symbol somehow *must* be given.

A solution to the specificity-problem is suggested by making a distinction between "occurrent images" and "dispositional (potential) images" (see Price, 1969). If, for instance, we attempt to think about dogs in general and form an image of a black dog, we are not misled into believing that all dogs are black, since we have other images (say, of a white dog) in stock which may supplement and generalize the original image. But now we must ask what guides the supply of relevant images. The answer seems inescapable that we must somehow possess the abstract concept in the first place.

Where the ambiguity of images is concerned, it is suggested that the less vivid and detailed, and the more "general" and "schematic" the image is, the more easily it can fulfill its conceptual representational function. But how is such a "schematic" image formed? The world contains no schematic dogs! Thus, while it is certainly true that we can think about something quite general and abstract when imaging, it seems clear that the image does not have a source of its own to provide the conceptual content of our thoughts.

In view of the serious shortcomings of the imagist theory, it is amazing how many of its intellectual genes have been active in shaping the form of our generation's imagery theories. Some of the more influential ones will now be reviewed. With the primary aim of accounting for cognitive development, Bruner (1964, 1966) has treated imagery wihin a general theory of symbolic representation in cognition. Although Bruner clearly transcends the general empiricist theory of knowledge (e.g., his treatment of language as a cognitive instrument), his concept of iconic representation has many features in common with the orthodox imagist theory. In the first place, images are seen to symbolize by way of *resemblance* to the objects they represent. Consequently, images are held to be closely linked to the *surface attributes* of things and are *concrete* and *static*. Secondly, imagery is seen as developmentally *prior* to language, and Bruner thus seems to think that imagery has a sort of bedrock function in providing a necessary substrate of knowledge for language to be founded upon. Imagery representations, then, function literally before language enters into the picture. No underlying representational code of a more abstract nature is specified, which could control the imagery function. To Bruner, then, images seem to be mental *pictures*, and here he seems to come close to one of the fatal ideas of the imagist theory. This is the notion that an image can "stand on its own feet" in cognition. According to Bruner's theory, there is a rather long period in the child's life where thinking mainly occurs in pictures, before language gets internalized and makes it possible to mount to higher and more advanced levels of thought. But, as already suggested above, this is a very questionable idea. The problem may be illustrated by way of an example presented by Fodor (1976). How could a sentence like "John is fat" be represented in a picture? A picture of John with a bulging tummy? If so, the picture representation does not distinguish the thought that "John is tall" or that "John is sitting, standing or lying," or simply "John" for that matter. The conclusion must be that a picture cannot "speak for itself," and that images, consequently, are always *images under description*. A picture, then, cannot "stand alone" as something that is "self-understanding."

The major stimulus for the renewed interest in the nature of imagery and its functional role in cognition undoubtedly came from the monumental research contribution of Allan Paivio (1971). In Paivio's conceptual scheme, both imagery and verbal processes are given the status of theoretical constructs within the framework of a neo-behavioristic mediational model. Paivio assumes a distinction between two symbolic codes - a verbal and a nonverbal imagery system. Although fundamentally independent, the two systems are assumed to be linked through implicit associative connections (a word may elicit an image and vice versa). Paivio's theory has, no doubt, served as a fruitful working model, and it has the merit of explaining a large number of interesting imagery effects in cognition. But his theoretical scheme also suffers from several serious weaknesses which limits its predictive and explanatory power. Here we will concentrate on some of the more important *conceptual* difficulties inherent in the model (see Kaufmann, 1980, pp. 43-46, for a more comprehensive critical discussion of Paivio's theory). The conceptual platform of Paivio's model has been taken over in surprisingly intact form from the orthodox imagist theory. The image is seen to correspond rather directly to the thing it represents, and symbolizes by way of resemblance to the thing it "stands for." The image is conceived, then, to be directly derived from perceptual experience, and the external events are held to be efficient causes of the resulting representation. As Yuille and Catchpole (1977) observe, no further mechanisms are required on Paivio's account of the acquisition of knowledge. Paivio's basic empiricist position is clearly exposed in his view on language acquisition, where he claims that "The major implication of the imagery approach for language acquisition is that linguistic competence and linguistic performance may be initially dependent upon a substrate of imagery. Through exposure to concrete objects and events, the infant develops a storehouse of images which represents his knowledge of the world. Language builds upon this foundation and remains interlocked with it, although it also develops a partly autonomous structure of its own" (Paivio, 1972,p. 29).

This is a clear commitment to the sense datum theory rejected above, and Paivio's position clearly implies that we have a direct and immediate access to the objects of an independently given world. Knowledge is seen to be literally contained in images, which are the primary symbols. Later on, the secondary symbols of language are added to this foundation.

A corollary of this position is to regard the image as a *dominant component of meaning* (see also Bugelski, 1977, 1982). But now we are up against serious problems. As argued above, the image cannot serve any basic meaning function, since the image itself has to be interpreted. Consequently, there must be something basically wrong with this kind of theory.

It is true that Paivio has argued against such criticism, by claiming that it is directed against a straw man. Since percepts are organized, selected and interpreted, why should it be so difficult to accept that the analog representations in imagery are interpreted too (Paivio, 1976)? But this line of reasoning offers no rebuttal at all of the argument in question; it only pushes the problem one step further away: Now we are left with the question of how percepts are interpreted!

Recently, Anderson (1978) has defended the dual code theory

against charges of internal inconsistencies. According to Anderson, images may be interpreted by being linked to words. Taken on its own, this argument has substance to it. But as a defense of Paivio's theory it is ill-founded. Now we must answer the question of how words gain meaning and are interpretable. Here Paivio would have to point to the image as the basic meaning substrate. But now we are in a sort of "semantic loop" where it is impossible to escape. Certainly, one cannot explain the meaning of an image by reference to a word, and then interpret the meaning of the word by way of the image.

The conclusion seems inevitable that there are serious shortcomings in Paivio's theoretical scheme. Problems also crop up when it comes to the *conceptual* dimension of the image. Paivio (1971) asserts that images may be schematic and abstract, and, in fact, he criticizes Bruner for not acknowledging this possibility. It is certainly true that when we have an image, it may mean something highly abstract to us. But is this development of imagery into schematic forms something that can occur *within* the image? Since it seems that Paivio endorses the classical distinction between images and words in the primary/secondary dimension, an answer somehow must be given to this question. And, as we have pointed out, no convincing answer is likely to present itself.

Since Paivio claims that the two symbolic systems are fundamentally *distinct*, we may wind up our discussion by asking how it is possible to think in a fundamentally nondiscursive, pictorial medium at all. Since any picture is indeterminate in its semantic/conceptual content, how are we to avoid getting utterly confused when thinking in a pictorial representational medium? It is a step forward to concede that images and words are units in a flexible system which can recombine, rearrange, and manipulate the units in the service of cognitive demands (Paivio, 1975a). The inevitable next step is to work out a basic revision of the theory, where such features are taken care of in a convincing way. As it stands today, the body of the theory obviously needs some new and vital organs that may secure its survival value.

Recently, Paivio (e.g., Paivio & Begg, 1981) has made an attempt to handle these problems by way of postulating a distinction between an *imagery system* and the *conscious imagery* itself. The imagery system is characterized as a system "which somehow retrieves or generates conscious images as well as other external manifestations and the conscious imagery itself. The system can function without necessarily producing conscious imagery, although such images can occur if necessary" (Paivio & Begg, 1981, p. 114). This maneuver seems to us to be unsatisfactory for several reasons. (1) It is intolerably vague, and no specification whatsoever is given as to the nature of the postulated imagery system beyond the point that it is somehow responsible for the generation of images. (2) It invites the slippery concept of unconscious imagery. One of the disturbing things here is that in operating with such an extended concept of imagery, a protective belt may be laid around the theory and, thus, immunize it against criticism based on conceptual considerations as well as on empirical findings. The problem relating to the existence of abstract and conceptual images may be "solved" by reference to the ill-defined and elusive "imagery-system." Negative empirical findings may be "explained" by postulating unconscious imagery process. (3) Moreover, if it is true that images are constructed under intentional descriptions and can "stand for" abstract ideas, as Paivio admits (ibid, p. 114), it must mean that the imagery system is basically a *propositional* representational system. The conclusion seems inescapable, then, that the postulation of hypothetical representational units in the form of "logogens" and "images" that give birth to consciously experienced verbal and imaginal representations commits Paivio to the acceptance of two additional systems of representations of a propositional nature - one underlying linguistic representations and another underlying imaginal representations. This is, perhaps, a possible solution, but it is not an attractive one, and it is inconsistent with Paivio's insistence on maximal parsimony in theory formulation (see, for example, Paivio, 1983). It also leaves unsolved the problem of how to *translate* between the different representational systems.

The other major theory of the analog variant is the one developed by Kosslyn (Kosslyn, 1980, 1981; Kosslyn, Pinker, Smith, & Shwartz, 1979; Pinker & Kosslyn, 1983). According to Kosslyn, visual images are most properly conceived as temporary spatial "displays" in an active memory. On Kosslyn's account, the image is a quasi-pictorial entity that *depicts* information, in contrast to *describing* it discursively. Thus, an image stands in a *nonarbitrary* relation to the object it represents. In the words of Kosslyn et al. (1979) ". . . . .any part of that representation is a representation of the corresponding part of the object" (p. 536). The image, then, symbolizes by way of resemblance to the object it represents. On this point Kosslyn stands on the same empiricist foundation as does Paivio. Kosslyn departs, however, from Paivio's conceptualization when he claims that the "surface representations" of imagery are generated from a general and abstract system of deep structure representations in long-term memory.

Since images are seen to be anchored in more abstract, conceptual representations, the essential property of the intentionality of images seems to have been taken care of. In light of this, Kosslyn takes a surprising turn in his theorizing when he insists on the existence and need for a separate interpretative mechanism (a "mind's eye") that works over ("looks at") the surface image with the purpose of identifying its various properties and sorting them into the proper semantic categories. In our view, this move shows how strongly wedded Kosslyn is to the traditional picture-in-the-mind theory despite his insistence to the contrary. Why should we have to "inspect" and iterpret an image that is constructed from an underlying conceptual representation? This strange idea leads Kosslyn to the even stranger notion that the image *itself* has (rather than represents) spatial properties (extent, form, etc.) that can be measured. A large part of Kosslyn's research program seems to be aimed at uncovering the spatial properties of the image. Here we believe that Kosslyn commits the fundamental error of confusing the image with the object it represents. As Hebb (1980) and Pylyshyn (1981) have pointed out, it is important to realize that when we have a visual image of an object, the spatial properties we "see" are properties of the object represented and *not* properties of the image *itself*. When we have an image of a large tomato, it is the object represented that is large, round and red, not the image! While Kosslyn seems to be aware of these problems, the notion of the image as an object to be "inspected" by the "mind's eye" invites just this idea, and Kosslyn explicitly endorses the view that images do have spatial extension. In our view, much of the experimental evidence presented by Kosslyn can simply (and trivially) be explained by assuming that his subjects interpret the tasks presented to them as requesting that they are to imagine (or simulate through imagery) what would happen under real perceptual conditions in relation to a situation specified by the experimenter. Essentially this point has been made by Pylyshyn (1981), and we think it is entirely convincing.

In the so-called "scanning" experiments, the subjects are asked to form an image of a scene (e.g., a map), which contains different objects at different positions. The subject is instructed to focus on a particular location of the image, and is then asked to verify the existence of different objects at different distances from the focal point of the image. The results show that the decision time increases systematically with the distance to be "scanned." The results are easily explained under the very likely assumption that the subjects take their task to be that of simulating what would happen under real perceptual conditions.

Kosslyn does not seem to appreciate the nature and force of this argument, as seems to be apparent from his counter-claim that the distance effect only appears with tasks independently judged to involve imagery, and by subjects reporting that they did use imagery (Kosslyn, 1981). But the argument against Kosslyn is just that when people *are* using imagery in these tasks, they do this with the purpose of ascertaining what would happen under real perceptual conditions. Kosslyn also refers to an experiment by Finke and Kosslyn (1980) where the subjects were asked to image pairs of dots moving toward the periphery and to report when the two dots were no longer distinct. A control group was included and informed about the instructions given to the experimental subjects. The task of the control group was to make the same judgment and they were explicitly told *not* to use imagery. The control subjects were only partly successful in their judgments as to what happened in the imagery condition. But this may be taken to mean only that using imagery is a convenient way of judging what happens under ordinary perceptual conditions, and that this is one of the important functions imagery brings to bear in cognition. In fairness to Kosslyn, the results of such experiments indicate that imaginal performance is truly functional and cannot be completely reduced to the utilization of tacit knowledge through a purely propositional representation, as Pylyshyn (1981) seems to believe. In other experiments, Kosslyn asks his subjects to form small or large images of objects (e.g., a cat), and claims that the results showing that it takes longer time to verify the existence of properties of small images substantiate the postulated quasi-pictorial properties of images, with the implication that small images are more difficult to "inspect" than larger ones. What these results may be taken to mean, however, is that more details are brought forward in the construction of large images, which is validated by the finding that it takes longer time to construct larger images. Kosslyn attempts to control for this alternative explanation, by comparing the relationship between decision time and association strength of a property to the noun of the object. He claims that the fact that the results showing that decision time is systematically related to the size and not to the association strength of a property excludes the alternative interpretation (e.g., decision time is longer for "cat claws" than for "cat head," where the former property is more closely associated with the noun). But the comparison with association strength to nouns is not very relevant in the present context. A more relevant comparison would be between decision time and the association strength of a property when the subject is drawing a picture of the object in question. It is reasonable to assume that subjects under imagery instructions are more likely to generate the property "cat head" than the property of "cat claws" for the reason that imagining is more like drawing a picture then seeing a picture.

In a different series of experiments, the subjects are asked to image an object as if it is seen from a long distance and then to imagine that they are gradually moving closer to the object. It is suggested that, at some point, the image will "overflow." At this point, the subject is to "stop" and estimate how far the object seems to be. The results of these experiments show consistently that smaller objects seemed to overflow at nearer distances than larger ones. From these findings, Kosslyn goes on to make some rather extraordinary suggestions. Kosslyn claims to have been able to measure the visual angle of the mind's eye, and, furthermore, that the imagey field seems to be round, which is held somehow to be a possible intrinsic property of the image. Again, the results may be interpreted in a less dramatic way. What is shown may only be that people are capable of simulating real physical events under ordinary perceptual conditions. A convenient way of solving the task set by the experimenter would be to simulate what would happen when one looks at the objects at the various distances through a lens and "zooms in" on the objects. The roundness of the imagery field may thus reflect the fact that the lens is round, and not that the image *itself* is round! The basic flaw in Kosslyn's thinking, then, seems to be that he confuses the properties of the image with the properties of the objects and scenes imagined. Since the theory is implemented in a computer program, it should effectively block arguments as to the incoherence of the notion of "mind's eye" operations interpreting a prior and independently formed image. Such a claim has, in point of fact, been made by Kosslyn et al. (1979). However, Kosslyn's theorizing here goes far beyond his computer model. The procedure mainly responsible for the interpretative work, FIND, is, in fact a *common* procedure. LOOK FOR is an inspection routine having the function of activating IMAGE, which is a generative procedure. There is, thus, simply no separate procedure which has the function of interpreting a prior and independently formed image. The computer model, then, covers, only what we think is the only defensible view, which is that interpretation and construction are strictly commensurate procedures. This is not to deny the important point made by Richardson (1980) to the effect that an image may not be determinate only under the interpretation that it was originally constructed. But this is an entirely different matter. The image is typically a *dense* representation that may *give rise to* new interpretations. This may be one of the reasons why imagery seems to be such an important working-space in the search processes and restructuring activity needed in creative thinking (see, e.g., Kaufmann, 1980, 1984). Kosslyn presents a wealth of interesting observations, such as piecemeal versus wholistic construction, the influence of verbal descriptions on imagery construction, the role of imagery in cognitive development etc. But no evidence whatsoever has been presented that should tempt us to believe that the image itself has spatial properties. We should not regret this, because, as Squires (1968) has argued, if we were looking at an object when imaging it in the same way as when perceiving, there could be no room for the distinct experience of visualizing!

A regrettable feature of Kosslyn's theory is that it contains virtually no significant clues as to the role of imagery in a broader cognitive context. What are the functional properties of imagery; that is, what are the "cognitive jobs" that are best executed by imagery in tasks of learning, memory and problem solving? In these respects, it seems to us that Paivio's approach has proved much more fruitful. The theory has stimulated an extensive research program that has resulted in a wealth of interesting and replicable findings on the effects of imagery in a wide range of tasks. The challenge is to construct a theory that can serve to integrate these findings under general and unifying principles. Kosslyn's theory clearly does not meet this challenge. It does not even address the relevant data. Kosslyn's position here is that before one can even begin to answer the question as to the functional utilities of imagery, the "nature" of imagery must be clarified. But a good argument can be made for the thesis that the nature of imagery can be clarified to a considerable extent through investigating its functions. At least, this should be the proper domain of psychological inquiry (see Kolers & Smythe, 1979), rather than trying to bring experimental evidence to bear on questions that are primarily conceptual in nature.

The view of imagery that emanates from Kosslyn's work is a severely restricted one. The scope of imagery is not limited to that of representing the physical properties of the real world. Images can be bent and twisted in all sorts of ways and have a corresponding multiplicity of functions to perform. As Arieti (1976) points out, the image can be used to represent not only the real but also the unreal. As such, it may constitute a "force of transcendence" (Arieti, 1976, p. 49). This is a big phrase and invites questions that are not easy to answer. Still these questions are real questions, and point to a genuine and important limitation in Kosslyn's theorizing about imagery.

III.PROPOSITIONAL REPRESENTATION AND THE CONCEPTUAL THEORY

A radically different account of the nature of imagery draws its main intellectual inspiration from what may be termed the "conceptualist theory," which is mainly rooted in rationalist epistemology (see Price, 1969). According to the conceptualist, the basic work in cognition is not mediated by either images or words. Rather, thinking is held to be a *unique* type of cognitive activity which may accompany and be expressed in language or imagery, and which may also occur in the *absence* of these. The proper substance of thinking is made up of mental entities of a special and abstract sort, variously described as "concepts," "abstract ideas," "propositional structures," "schemata," etc. Imagery is thus placed in a purely external and adventitious relation to the act of thinking and is either regarded as epiphenomenal or assigned a secondary, auxiliary function.

The conceptualist theory was brought into experimental psychology by the members of the Wurzburg school (see Mandler & Mandler, 1964). In their kernel idea of "imageless thought," thinking is seen as a fresh mental category that does not require representation through any kind of symbolic particulars (i.e., images and words). The basic elements of thought are "abstract ideas," rather than concrete images.

Within the conceptualist scheme there is no problem in accounting for the semantic/conceptual content of images, since images are seen as the *products* of more basic, abstract mental processes which are conceptual in nature. It also escapes the problem of identifying thinking with the mere occurrence or production of symbols in "full dress," which burdens a mediational theory of the Paivio-type. After all, it is extremely rare for thinking to be constantly formulated in words, in a kind of well-articulated inner speech. Nor does it happen very often that thinking occurs through a constant stream of full-fledged images.

This does not, of course, prove the conceptualist theory to be correct. After all the notion of thinking as a "unique," "impalpable" experience is rather abstruse, and the Wurzburg theorists never managed to develop it in any constructive way.

Let us now examine the contemporary variants of the conceptualist theory. Here we may distinguish between "hard-nosed" versions, where the image is seen as epiphenomenal, and "soft-nosed" versons, where imagery is assigned a secondary, supportive function in cognition.

An articular defense of the hard-nosed version is provided by Pylyshyn (1973) in his comprehensive and thorough criticism of the systematic status assigned to imagery in mediational models. According to Pylyshyn, images are pre-interpreted, selected and may be abstract, and the general notion that "all learning and memory - and, indeed, all cognition - takes place exclusively through the medium of either words or images" (ibid, p. 4) is totally misguided. Rather, cognition is mediated by "something quite different" from either words or images. This something else is variously described as "concepts," "abstract mental structures" or "propositional structures," which refers to an abstract and amodal representational format of a purely conceptual nature. The existence of images is, of course, not denied, but they are regarded as epiphenomenal ripplings on the surface, devoid of functional value. As far as language is concerned, it is mainly assigned a labeling function in thought, and is conceived primarily as an instrument for communication of the preformed nonverbal mental concepts (p. 7). The close kinship of Pylyshyn's views to the orthodox conceptualist theory should not be difficult to discern.

To back up his claim, Pylyshyn seeks support in the work of the philosopher, Frege (1879/1960). From the example of two sentences which express the same thought, Frege makes the following comment which Pylyshyn takes to his heart: "..... I call the part of the content that is the same in both *the conceptual content. Only this* has significance for our symbolic language; we need therefore make no distinction between propositions that have the same conceptual content" (ibid, p. 3). For Pylyshyn this is tantamount to saying that the concept in question must be a *mental* entity. Since we can express the same thought in different sentences, there must be an underlying mental entity of a different nature than the words used to express it. But here Pylyshyn mistakes this prosecutor for his defense lawyer. Frege explicitly opposed, and, indeed, rebelled against the identification of a concept with a mental entity. For Frege, the concept in question was an *abstract* entity and *not* a mental entity (see also Putnam, 1975). The sentences may express the same thought in much the same way as two different instruments may be brought to perform the same function. The direct move from the existence of abstract entities to the necessary existence of linguistically neutral abstract *mental* entities existing in an elusive conceptual medium at the deep structure level of the mind, is, thus, not legitimate. We believe that the temptation to make this move from the considerations that Pylyshyn expresses is a primary motive for postulating the sort of abstract, amodal representational system that has won such appeal in current cognitive theories (see Anderson, 1976, 1980; Anderson & Bower, 1973; Kintsch, 1974; Lindsay & Norman, 1977; and Norman & Rumelhart, 1975). We regard this an unfortunate move, because, as have been pointed out by Harman (1975), it only pushes the problem back one step (see also Kosslyn & Pomerantz, 1977). Now we have to explain how meaning is established in the propositional representations. We agree with Harman when he claims that "One cannot continue forever to translate one system of representation into another," and that this kind of theory "..... only delays the moment of confrontation" (Harman, 1975, p. 282), where we realize that a different account is needed. The only way to save this kind of theory seems to be that the underlying conceptual language that we are supposed to think in is, somehow *intrinsically intelligible*. This solution has been advocated by Fodor (1976), who claims that our basic conceptual knowledge (left unspecified) is *innate*. But the idea that such concepts as "electron" and "xylophon" are innate hardly seems an attractive one. (For a comprehensive critical discussion of Fodor's theory of the language of thought, see Kaufmann, 1980).

Pylyshyn has another card in his sleeve that must be considered. The existence of a propositional representation is held to be necessary to explain other problems related to the meaning function. Claims Pylyshyn: "As long as we recognize that people can go from mental pictures to mental words or vice versa, we are forced to conclude that there must be a representation (which is more abstract and not available to conscious experience) which encompasses both. There must in other words be some common format or interlingua" (ibid p. 5).

But nothing forces us to follow Pylyshyn in this assumption. On the contrary, since the underlying propositional representational units are regarded as fundamentally different from words and images, it can be shown that Pylyshyn's argument is logically incoherent. To be able to translate from Code 1 to Code 2, it is claimed that we have to go via an underlying Code 3. But to translate from Code 1 to Code 3, we must postulate a Code 4, etc. An infinite regress has been generated and nothing has been explained by resorting to an abstract, underlying metalanguage (see also Kosslyn & Pomerantz, 1977).

Anderson (1978), who favors the propositional theory of representation, nevertheless accepts the validity of this argument. The conclusion he draws is, however, surprising indeed: "It is simply not the case that it is necessary to have a propositional or any other intermediate code for translation. By careful analysis, it might be possible to show that an interlingua makes the translation process more efficient, but such an analysis has not been forthcoming" (p. 256). But what the argument shows is not that an interlingua is not *necessary* to effect the translation, but rather that it is *not possible* to solve the translation problem by postulating an interlingua at a deep structure level, which is fundamentally different from words and images. When an infinite regress is generated, it means that the idea in question is logically corrupt. The translation problem is, in a sense, real enough, and a sound theory of representation has to offer a convincing solution to it. None of the theories we have considered so far, however, seem fit to cope with the problem.

Disregarding such conceptual problems as those outlined above, and viewing matters in more strictly empirical terms, it also seems that the hard-nosed theory runs into serious difficulties in explaining imagery effects in cognitive performances, such as modality specific interference effects (Atwood, 1971; Brooks, 1967; Byrne, 1974; Salthouse, 1974, 1975), memory for modality information (see Kieras, 1978, for a general review), the mental rotation phenomenon (Shepard, 1978a, 1978b) and many other phenomena (see Kosslyn & Pomerantz, 1977 and Paivio, 1975a, 1975b for detailed discussions of the empirical side of the issue). Although Anderson (1978) has been clever in pointing out how the propositional model may accommodate many of these findings, we agree with Hayes-Roth (1979) that the propositional account is notoriously ad hoc, and that the imagist theories gain the upper hand in having predicted the results. This upper hand is conditional, of course, on the assumption that these results are not determined artefactually by the same tacit theories and knowledge held by the subjects as those shared by the investigators. This general state of affairs in the field has consequently fostered compromise variants of the propositional theory, where imagery is given a functionally autonomous role in cognition. As an example, we may choose the model advocated by Yuille and Catchpole (1977), which is mainly inspired by Piaget's theory of symbolic representation.

According to Yuille and Catchpole we have to assume the existence of two *distinct* levels of cognitive functioning which are referred to as the *representational* and the *abstract* planes. Since, in their view, "..... images and words are too concrete to serve as the basis for most cognitive operations" (p. 177), the fundamental cognitive processes' are done "off-stage" at the abstract level "in a form and symbolism unique to the mind" (p. 177). Basic knowledge is also seen to be stored in the abstract coding system. Imagery is located at the representational level and assigned several supportive functions, such as the temporary maintenance of a constructed representation, reconstruction of ideas, memories, feelings, etc. from the abstract code, as well as making possible a more direct recovery of past experience.

Such a hybrid model has the merits of attempting to deal constructively with both the semantic/perceptual and the sensoric/perceptual aspects of imagery. Unfortunately, this liberalization of the propositional theory does not solve its basic problems. Since we are dealing with distinct levels in cognition where the basic processing occurs at the abstract level in a representational system that is different from words and images it follows, as Yuille and Catchpole (1977) point out themselves, that "..... the contents and activities of the abstract plane are not available to conscious inspection ..... conscious inspection of an abstract idea requires its translation into a representational mode" (p. 179). In other words, our thoughts are hatched out at a deep-structure level in a unique "language of thought" beyond conscious access. Then, somehow, they are translated into a concrete representation (in images or words) and made accessible to consciousness.

In our opinion, this is a highly problematic implication of the model. In the first place, such dissection of thinking into two distinct spheres (generation and representation) really implies that our thoughts should constitute a continuous source of potential surprise to us. There are elements of this feature in our mental experience, particularly in the case of "unbidden images." However, since this is normally not the case, there must be something fundamentally wrong with the ontological division between the abstract and the representational level.

Apart from such perplexing implications of the theory, a logical impasse blocks the road from the abstract to the representational level. Since these two levels are distinct, and the abstract, underlying representational system is different from words and images, we have to postulate a sort of "mental link station," that is, the two codes must be interconnected by a translator-code that somehow encompasses both. But to move from the abstract code to the translator code, we should need a new translator code, and so on. Again we are involved in an infinite regress and there seems to be no way of reaching the representational level from the abstract code.

No matter how attractive the cognitive theories may seem in many important respects, the conclusion nevertheless seems inevitable that there is something fundamentally wrong with their conceptual foundation.

So it seems that we have to continue our search for a sound conceptual basis for a theory of symbolic representation. Let us, therefore, turn to the third major group of imagery theories.

IV. PERCEPTUAL ANTICIPATION AND DISPOSITIONALISM

A quite different conceptual approach to imagery is to conceive an image as a *disposition to see the thing imagined*. Ryle (1949) may be interpreted as advocating a dispositional analysis of imaging. Ryle admits, however, that such an account cannot properly be applied to thoughts, feelings and sensations, and imagery may be seen as belonging primarily in this conceptual family (see Hannay, 1971 for a comprehensive exposition and critical discussion of Ryle's theory).

Recently, Neisser (1976) has presented a theory of imagery which is clearly in this tradition. Neisser rejects both the analog theory and the propositional theory of imagery. The analog theory is said to entail serious internal inconsistencies (e.g., implying undischarged homunculi). Furthermore, both theories are claimed to be unable to explain basic facts about imagery (e.g., why images and percepts are not systematically confused).

In trying to avoid the alleged difficulties of his theoretical opponents, Neisser "depictorializes" imagery and adopts the Rylean strategy of reducing the image to a dispositional status. According to Neisser (1978), the true nature of the image is captured in regarding it as a *perceptual anticipation*, that is, when imaging we are "preparing for exterospection" (p. 173). The very essence of imagery, then, is perceptual readiness."To have a perceptual set for something is to have an image. The more precisely that image anticipates the information to come, the more effective the set should be" (Neisser, 1976, p. 145).

But this conception of imagery is problematic in many respects. Whereas a limited class of imaging may, perhaps, be validly described as perceptual anticipations, it seems to us that in the normal case of imagery it is quite the *opposite* that is happening: When we are imaging, we do it rather as a *substitute* for real perceptual experience and precisely because we do not anticipate any "exterospection." Suppose that we were to describe the looks of a deceased relative of ours. Since we certainly cannot expect any perceptual input we may summon up an image of the person to aid us in our task. Neisser is aware of this problem, but his attempt to show that it makes sense to speak of anticipations even in the cases where we do not anticipate anything is simply not successful.

Matching the grammar of the concept of perceptual anticipation with the grammar of the concept of imaging, we clearly see that they are fundamentally different phenomena. In a thoroughly justified criticism of Neisser's theory, Hampson and Morris (1978) refer to a critique by Hannay (1971) of Ryle's dispositional account of imaging, which transfers with full force to Neisser's theory. Where a perceptual anticipation goes unfulfilled, according to Hannay, we always experience surprise or disappointment. Such description does not fit images in the same way. Consequently, images cannot all be perceptual anticipations.

In reply to Hampson and Morris, Neisser (1978) claims that this is demonstrably wrong, since many kinds of anticipations are unfulfilled without surprising or disappointing anyone. To substantiate this claim, Neisser refers to the example of a seed which is described as "..... a highly structured set of anticipations - it is ready for the warmth and nutrients that will enable it to grow - but no one supposes that seeds are capable of surprise" (p. 191). Of course, no one supposes that seeds are capable of surprise, but that is because no one should suppose that seeds are capable of having anticipations in the first place! Speaking of a seed as having "anticipations" is just as misleading and erroneous as describing a magnet as "recognizing" the magnetic meridian, or of the lymphocytes of the body as "remembering" a smallpox injection (the examples are borrowed from Toulmin, 1971, where a general discussion of the appropriate conditions for employing cognitive terms is found). In the case of the seed having "anticipation," the context of purposive behavior where it is appropriate to assign cognitive terms to the behavior of organisms is completely absent. (The same argument applies to another example of Neisser's where the military system is endowed with "anticipations.").

So Neisser has not been able to meet the serious challenge presented to him by Hampson and Morris. He merely circumvents it by stepping outside the logic of the concept of anticipation. We may conclude, then, that Neisser has placed imagery in the wrong logical category, and that the true essence of imagery is certainly not captured by regarding it as a perceptual anticipation. Rather than to clarify, Neisser's theory seems to seriously distort our view on the nature and functions of imagery.

V. AN ALTERNATIVE MODEL

Thus, it seems that none of the main theories escapes serious conceptual problems. Is there indeed any hope of finding a definite solution to these intricate problems? At least we may try, and in the following we will outline a conceptual underpinning which we believe is able to keep the theoretical structure upright. We will single out some of the assumptions believed to constitute the main roots of the trouble for existing theories. While we proceed, we will state our own position on these issues. Let us start with the more general assumptions.

1. The Mental Process Picture of Thinking

It may seem odd to question the assumption that thinking is a mental process. Everybody certainly agrees on *that*. In a certain sense, thinking does, of course, involve mental processes. Nevertheless, we believe that the different parties are misled into distorted theoretical positions by sticking to too close a parallel between a mental process and a physical process, and, furthermore, by too closely *identifying* thinking with a mental process.

More specifically we will attempt to defend the following claims: The imagist and dual-code theorists want to be able to point to a tangible process. Thus, they arrive at too close an identification of thinking with fully dressed-up symbolic processes in the form of strings of words and images mediating the connection between stimulus and response (e.g., Berlyne, 1965; Paivio, 1971).

The conceptualist-inclined theorist rightly rejects the identification of thinking with a mediational process of strings of words and images, but wrongly assumes from this that thinking must have a "nature of its own" in the sense that it is to be identified with an autonomous mental process operating at a deeper symbolic level, where the basic elements are "abstract meanings," "concepts," etc., which, now and then, "send up" a word or an image to keep the thinker informed, as it were. The radical behaviorist may also be regarded as tricked by the mental process picture: After failure to pin down the presumed mental process in a definite way, the existence of mental acts is denied altogether, and thinking is alternatively considered as hypothetical and semi-hypothetical statements about overt behavior. On this scheme the existence of images is either flatly denied, or images are conceived as dispositions to see the things imagined.

Let us now attempt to outline an alternative position which is constructive and does not deny any obvious feature of mental life. Mental acts *do* occur which *are* episodic (that is: which occupy a position in a time series and which are expressed in the symbolic codes of language and imagery). Mental acts should not, however, be construed as mental processes analogous to physical processes. Why not? As has been persuasively argued by Wittgenstein (1953) and Geach (1971), we cannot in *principle* assign a definite position in time to mental acts in the same way as to physical events. When attempting to get hold of a mental act in a definite way in physical time, a host of problems are whirled up. Does the mental act occur *in advance* of its expression? What, then, did the thought consist of before its expression? How can we give a constructive answer to such a question? Is the mental act, then, rather like a series-explosion triggered at different intervals with cognitive blank spots in between them which are filled by "surface" symbols with no mental charge in them? Or is the mental act simply going on continuously during its expression? In this case, we would have to say that a thought occurred at time "t" and lasted exactly as long as it took to express it. Such a description seems clearly absurd.

The view expressed here should, of course, not be taken to mean that mental acts occur in a mysterious, super-physical realm. It only means that mental acts are more loosely related to time than a physical process is. But, as Geach (1971) points out, that mental acts are loosely related to time "..... is still for them to be tied" (p. 106).

Where may the true solution lie then? As our general position, we claim that neither should thinking be identified with verbal and imaginal mediational process, nor with an autonomous mental process separate from its symbolic expression or formulation. Rather, we should conceive of thinking as *immanent and constituted* in its adequate symbolic expression (see Wittgenstein, 1953, for instructive examples of perplexities involved in trying to separate thinking from its expression).

Furthermore, we believe it is important to see that thinking takes place in a wider *situational context* which it is independent upon for its significance. The mental act, its symbolic expression and its appropriate context should, then, be regarded as forming an organic unity.

There seems to be two important points to notice as fas as the context is concerned: (1) First, there is what we may call the *intentional* aspect. As has been pointed out by many philosophers (e.g., Wittgenstein, 1953) and psychologists (e.g., Rommetveit, 1968, 1974), the pragmatic context that a symbolic expression occurs in is an important determinant of its significance, that is, what the expression means. The point here is primarily important in the context of communication. (2) Second, and most important for our present cocnern, is what we will call the *functional* aspect, which is primarily related to the context of thinking and problem solving. Symbolic processes have evolved to enable the individual to cope efficiently with environmental demands. It is, therefore, important to specify the constituents of the task environment that make it "unstructured" and give rise to uncertainty. Such factors include novelty, complexity, inconsistency, etc., or what Berlyne (1965) called "collative variables." We should, then, examine systematically the efficiency of the symbolic systems in relation to such variables. In this way we may gain valuable knowledge as to the functional properties of the different symbolic systems (e.g., linguistic and imaginal) and of their role in cognition in the broad sense.

2. Meaning and Concepts as Mental Entities

A further source of confusion for the theories discussed above is the implicit assumption that meanings and concepts must be some sort of entity. The imagist and dual-code theorist point to the image as the central substance of meaning. But, as we have seen, the image cannot fulfill any basic meaning function. The conceptualist theorists treat meaning as some sort of abstract idea or propositional structure that may provide semantic life to otherwise dead words and pictures. But this move to the deeper abstract level of mind seems not to move us any further to a constructive solution, since it is by no means clear what is meant by such an "idea" or "abstract mental structure" and how its meaning function is established. Here we will follow the lead taken by Wittgenstein (1953). According to Wittgenstein, "meaning" does not stand for any entity at all. Rather, the term "meaning" is a circumscription for statements about the *use* of signs (see Specht, 1961, for a lucid exposition of Wittgenstein's theory of meaning). From the Wittgensteinian perspective, linguistic activity stands in no need of a mental correlate in the form of images or abstract mental ideas or processes, in order to be meaningful. Wittgenstein's analysis, however, may primarily be seen as relevant to the level of meaning related to communication and to the use of language in certain institutions, rituals and practice of a group of speakers (Harman, 1968). In the context of cognitive funtioning, we will suggest that the analysis of meaning in terms of *conceptual role* advocated by Harman (1975, 1977) may be the appropriate one. In this view, the meaning of a sign or sentence is its role in the system, or, more specifically, its role in a conceptual scheme. With reference to the work of the philosopher Wilfrid Sellars (1963), Harman identifies the meaning of a symbolic expression with its (*potential*) *role in the evidence-inference-action language game of thought*. Black (1972) makes a similar point and claims: "The 'life' of the words, we might say, is not in some supposed mental afflatus, but rather in the capacity of the particular utterance to interact with, and provide a point of departure for further symbolic activity" (p. 30). So rather than referring to a mental entity, in the form of an image or an abstract idea, "meaning," in the psychological sense, refers to our *ability* to handle interrelated and coordinated functions of signs within a conceptual scheme. From these considerations it seems natural and promising to follow the line advocated by Harman (1970, 1975, 1977) leading to the view that the primary use of language is in *thought*, and that its use in communicaion is secondary and derived. In this view, learning a language is learning *a new way of thinking* which enable us to acquire a vast repertoire of propositional attitudes and conceptual resources that we would not have *without* language. Taking a step further, we will suggest that the acquisition of language is primarily the acquisition of a representational system that makes genuine conceptual and discursive thinking possible. On this scheme, the underlying conceptual representations used in cognition may be seen to consist in *linguistic* deep structures, or tokens of such deep structures. Thus, the abstractness and conceptual nature of the underlying representations in cognition may be accounted for by assuming that the representational units in question are (a) particular, meaning-strategic words which may be coupled with conceptually significant images, and (b) by way of the particular arrangements these representational units are organized in.

3. Images as Perceputal Analogs or Symbolic Descriptions

Zooming in on the more specific imagery assumptions, we note that there is a polarization in the views pertaining to the representational *content* of the image.

The analog theorist is too impressed by the sensoric-perceptual nature of the image to appreciate the significance of its semantic/conceptual aspect. On the other hand, the propositional and the anticipation theorists are so impressed by the semantic-conceptual side of the image that the sensoric-perceptual dimension is ignored.

But there is no need to polarize the discussion into a question of whether images represent in a pictorial *or* descriptive way. They do both. Fodor suggests that there may be an ". . . . indefinite range of cases in between photographs and paragraphs" (Fodor, 1981, p. 144). Marks (1983) argues that there is a continuum from the lowest levels which produces vivid quasi-perceptual imagery to the highest levels where imagery is abstract and conceptual. With these views we agree and emphasize that imagery may have important functions in cognition by virtue of its pictorial quasi-perceptual content.

Let us draw some preliminary conclusions: Our discussion seems to have made it quite clear that images cannot "stand on their own," as pictures. A picture has no semantic/conceptual source of its own, nor has it the needed discursive properties to function as a vehicle for thinking. Imagery has, thus, to function within and under control of a more superordinate representational system which does have these properties.

We have seen that the idea of an abstract conceptual representational system which is different from linguistic and imaginal representations and controls peripheral codes (verbal and visual) leads us into serious logical incoherences. Neither do we need such an abstract coding system. Language possesses all the properties that are necessary in order to account for the discursive properties and semantic-perceptual nature of thinking. We may now identify the general place of imagery in cognition: It is an *ancillary representational system operating mainly within and under conceptual control of the linguistic symbolic system.* In this place imagery can execute important functions (see e.g., Kaufmann, 1980, 1984).

Does this mean that thought is impossible without language? It does not! Here we must draw a distinction between *explicit* and *implicit* representation. Explicit representation enables us to think in a discursive way on a semantic-conceptual level and is largely dependent on our ability to use language as a cognitive instrument, supported by the symbolic system of imagery. This we hold to be a unique capacity of humans, which makes it possible for us to stand back from the immediate stimulation from the environment in a detached sort of way and achieve the "generalized reflection of reality" that Vygotsky (1962) speaks of.

Thinking may also, however, be actualized through implicit representation in terms of functional organization. That is, thinking may be represented implicitly in the form of action programs tied to and activated in more specific contexts.

It may well be the case that some rudimentary and primitive pre-linguistic imagery representation may have a functional place in an intermediate zone between implicit and explicit representation. Such pre-linguistic imagery processes could have the anticipatory funtion that

Neisser (1976) suggests, where the image is a preparation for some expected perceptual event. The symbolic significance of such imagery may be seen to lie in its functional role in an anticipatory scheme. It is the deliberate and purposeful use of imagery under explicit representation that in our theory has to be seen as being organized *within* the superordinate linguistic representational system. It is interesting to note that a growing body of evidence suggests that the usefulness of imagery as a mediational strategy *increases* with age (e.g., Kosslyn, 1980; Levin & Pressley, 1978). Such findings fit nicely with the theoretical scheme suggested here and present serious problems for other theories that address themselves to the developmental side of symbolic representation in cognition (e.g., Bruner et al., 1966; Paivio, 1971).

It is now time to examine our theory in relation to the basic conceptual issues discussed above: (1) Since, in our theory, the "inner" language that we think in incorporates the "outer" language that we speak in, and thinking is held to be internally related to its symbolic expression, no difficulty arises in translating from the "abstract" to the "representational" level. Also avoided are the problems involved in *identifying* thinking with verbal-imaginal mediational processes, or with an autonomous mental process at an abstract level. (2) Since imagery is seen to be nested into language as a subsidiary representational system, no difficulties are involved in accounting for its semantic-conceptual content, and the "inter-code" translation problem is also handled. (3) The theory recognizes the quasi-perceptual aspect of imagery, and its potential functional value in cognition.

On the empirical side the theory entails that imagery has the special utility of making possible an *increased amount of processing* in cognition (see Kaufmann, 1980, 1984). More specifically, the functions assigned to imagery may give improved possibilities for *elaboration* of information, which may be of particular significance in learning and memory. Furthermore, imagery may be particularly suited for embodying *mental models* for search processes and restructuring activity in problem solving when level of programming in the task environment is low (see Kaufmann & Helstrup, 1985).

VI. CONCLUDING REMARKS

It would be naive not to expect that important qualifications of the general suggestions made above will have to be made as research proceeds and we learn more about the micro-structure of imagery and cognition. As the research evidence stands today, however, it seems to be consistent with the notion of imagery as an ancillary symbolic tool, the use of which makes possible an increased amount of processing in cognition which enhances learning, memory and problem solving performance to a significant degree. Even if the hard work of determining precisely what kind of information processing functions imagery serves in cognition largely remains, we should first clarify the more general issues as to the role of imagery in cognition. A potentially important limitation of the present theoretical approach is that it represents a purely *cognitive* perspective. This may turn out to be too narrow a scope in dealing with the issue of mental representations. The recent intriguing research reported by Rogers (1983) on emotion as a "third code" points in this direction. Ahsen (1982), in his triple-code-theory formulation, also presents telling arguments against a purely cognitive perspective such as the one we have proposed.

REFERENCES

Ahsen, A. (1982). Imagery in perceputal learning and clinical application. *Journal of Mental Imagery, 6,* 157-168.

Anderson, J.R. (1976). *Language, memory and thought.* Hillsdale, NJ: Lawrence Erlbaum Associates.

Anderson, J.R. (1978). Arguments concerning representations for mental imagery. *Psychological Review, 4,* 249-277.

Anderson, J.R., & Bower, G.H. (1973). *Human associative memory.* Washington D.C.: Hemisphere Publishing Co.

Arieti, S. (1976). *Creativity.* New York: Basic Books.

Atwood, G. (1971). An experimental study of visual imagination and memory. *Cognitive Psychology, 2,* 290-299.

Baker, G.P., & Hacker, P.M.S. (1980). *Wittgenstein: Understanding and meaning.* Oxford: Basil Blackwell.

Baylor, G.W. (1972). *A treatise on the mind's eye.* Unpublished doctoral dissertation.

Berkeley, G. (1952). A treatise concerning the principles of human knowledge. (Original work published 1710). In R.M. Hutchins (Ed.), *Great books of the western world.* London: Encyclopedia Brittanica, Inc.

Black, M. (1972). *The labyrinth of language.* London: Pelican Books.

Block, N. (Ed.) (1981). *Readings in philosophy of psychology. Vol. 2.* London: Methuen.

Block, N. (Ed.). (1982). *Imagery.* Cambridge, MA: M.I.T. Press.

Boden, M.A. (1979). The computational metaphor in psychology. In N.Bolton (Ed.), *Philosophical problems in psychology.* London: Methuen.

Brooks, L.R. (1967). The suppression of visualization by reading. *Quarterly Journal of Experimental Psychology, 19,* 289-299.

Bruner, J.S. (1964). The course of cognitive growth. *American Psychologist, 19,* 1-15.

Bruner, J.S., Olver, R.R., Greenfield, P.M., et al. (1966). *Studies in cognitive growth.* New York: John Wiley & Sons.

Bugelski, B.R. (1977). Imagery and verbal behavior. *Journal of Mental Imagery, 1,* 39-52.

Bugelski, B.R. (1982). Learning and imagery. *Journal of Mental Imagery, 6,* 1-92.

Byrne, B. (1974). Item concreteness and spatial organization as predictors of visual imagery. *Memory & Cognition, 2,* 53-59.

Chase, W.G., & Clark, H.H. (1972). Mental operations in the comparison of sentences and pictures. In L.W. Gregg (Ed.), *Cognition in learning and memory.* New York: John Wiley & Sons.

Finke, R.A., & Kosslyn, S.M. (1980). Mental imagery acuity in the peripheral visual field. *Journal of Experimental Psychology: Human Perception and Performance, 6,* 126-139

Fodor, J.A. (1976).*The language of thought.* Hassocks, Sussex: Harvester Press.

Fodor, J.A. (1981). Imagistic representation. In N. Block (Ed.), *Readings in philosophy of psychology.* II. London: Methuen & Co.

Frege, G. (1879). Begriffschrift. In P. Geach & M. Black (Eds.). (1960). *Translations from the philosophical writings of Gottlob Frege.* Oxford: Blackwell.

Geach, P. (1971). *Mental acts.* London: Routledge & Kegan Paul.

Hampson, P.J., & Morris, P.E. (1978). Cyclical processing: A framework for imagery research. *Journal of Mental Imagery, 3,* 11-22.

Harman, G.H. (1968). Three levels of meaning. *The Journal of Philosophy, 65,* 590-602.

Harman, G.H. (1970). Language learning. *Nous, 1,* 33-43.

Harman, G.H. (1975). Language, thought and communication. In K. Gunderson (Ed.), *Language, mind and knowledge.* Minnesota Studies in the Philosophy of Science, Vol. VII, Minneapolis: Univeristy of Minnesota Press.

Harman, G.H. (1977). *Thought.* New Jersey: Princeton University Press.

Hayes-Roth, F. (1979). Distinguishing theories of representation: A critique of Anderson's "Arguments concerning representations for mental imagery." *Psychological Review, 4,* 376-382.

Horowitz, M.J. (1970). *Image formation and cognition.* London: Butterworths.

Kaufmann, G. (1980). *Imagery, language and cognition.* Oslo/Bergen/Tromso: Universitetsforlaget.

Kaufmann, G. (1984). Mental imagery and problem solving. *International Review of Mental Imagery, 1,* 23-57.

Kaufmann, G., & Helstrup, T. (1985). Mental imagery and problem solving: Implications for the educational process. In A.A. Sheikh & K.S. Sheikh (Eds.). *Imagery and education.* New York: Baywood Publishing Company.

Kintsch, W. (1974). *The representation of meaning in memory.* Hillsdale, NJ: Lawrence Erlbaum Associates.

Kolers, P.A., & Smythe, W.E. (1979). Images, symbols and skills. *Canadian Journal of Psychology, 33,* 158-184.

Kosslyn, S.M. (1980). *Image and mind.* Cambridge: Harvard University Press.

Kosslyn, S.M. (1981). The medium and the message in mental imagery: A theory. *Psychological Review, 88,* 46-66.

Kosslyn, S.M., & Pomerantz, J.R. (1977). Imagery, propositions, and the form of internal representational systems. *Cognitive Psychology,* 52-76.

Kosslyn, S.M., Pinker, S., Smith, G.E., & Shwartz, S.P. (1979). On the demystification of mental imagery. *The Behavioral and Brain Sciences, 2,* 535-581.

Lindsay, P.H., & Norman, D.A. (1977). *Human information processing.* New York: Academic Press.

Levin, J.B., & Pressley, M. (1978). A test of the developmental imagery hypothesis in children's associative learning. *Journal of Educational Psychology, 70*(5), 691-694.

Mandler, J.M., & Mandler, G. (Eds.). (1964). *Thinking: From association to gestalt.* New York: John Wiley & Sons.

Marks, D. (1977). Imagery and consciousness: A theoretical review from an individual differences perspective. *Journal of Mental Imagery, 2,* 275-290.

Marks, D. (1983). Mental imagery and consciousness: A theoretical review. In A.A. Sheikh (Ed.), *Imagery: Current theory, research and application.* New York: John Wiley & Sons.

Mehler, J., Walker, E.C.T., & Garrett, M. (Eds.). (1982). *Perspectives on mental representation.* Hillsdale, NJ: Lawrence Erlbaum.

Neisser, U. (1976). *Cognition and reality.* San Francisco: W.H. Wheeler & Co.

Neisser, U. (1978). Anticipations, images and introspection. *Cognition, 6,* 169-174.

Paivio, A. (1971). *Imagery and verbal processes.* New York: Holt, Rinehart & Winston.

Paivio, A. (1972). Imagery and language. In S.J. Segal (Ed.), *Imagery: Current cognitive approaches.* New York: Academic Press.

Paivio, A. (1975a). Neomentalism. *Canadian Journal of Psychology, 29,* 263-291.

Paivio, A. (1975b). Imagery and synchronic thinking. *Canadian Psycholgocial Review,* 147-163.

Paivio, A. (1976). Images, propositions, and knowledge. In J.M. Nicholas (Ed.), *Images, perception and knowledge.* (The Western Ontario Series in the Philosophy of Science). Dordrecht, The Netherlands: Reidel.

Paivio, A. (1983). The empirical case for dual coding. In J.C. Yuille (Ed.), *Imagery, memory and cognition.* Hillsdale, NJ: Lawrence Erlbaum.

Paivio, A., & Begg, I. (1981). *Psychology of language.* New Jersey: Prentice Hall, Inc.

Pinker, S., & Kosslyn, S.M. (1983). Theories of mental imagery. In A.A. Sheikh (Ed.), *Imagery: Current theory, research, and application.* New York: John Wiley & Sons.

Price, H.H. (1969). *Thinking and experience.* London: Hutchinson.

Putnam, H. (1975). The meaning of 'meaning.' In K. Gunderson (Ed.), *Language, mind and knowledge.* Minnesota Studies in the Philosophy of Science, Vol. VII, Minneapolis: University of Minnesota Press.

Pylyshyn, Z.W. (1979). Validating computational models. A Critique of Anderson's indeterminacy of representation claim. *Psychological Review, 4,* 383-394.

Rogers, T.B. (1983). Emotion, imagery and verbal codes: A closer look at an increasingly complex interaction. In J.C. Yuille (Ed.), *Imagery memory and cognition.* Hillsdale, NJ: Lawrence Erlbaum Associates, Inc.

Rommetveit, R. (1968). *Words, meanings and messages.* New York: Academic Press.

Rommetveit, R. (1974). *On message structure.* London: John Wiley & Sons.

Rorty, R. (1980). *Philosophy and the mirror of nature.* Oxford: Basil Blackwell.

Ryle, G. (1949). *The concept of mind.* London: Hutchinson.

Salthouse, T. (1974). Using selective interference to investigate spatial memory representation. *Memory and Cognition, 2,* 749-757.

Salthouse, T. (1975). Simultaneous processing of verbal and spatial memory representation. *Memory & Cognition, 3,* 221-225.

Sellars, W. (1963). *Science, perception and reality.* London: Routledge & Kegan Paul.

Shepard, R.N. (1978a). The mental image. *American Psychologist,* February, 125-137.

Shepard, R.N. (1978b). Externalization of imagery and the act at creation. In B.S. Randhawa (Ed.), *Visual learning, thinking and communication.* New York: Academic Press.

Specht, E.K. (1963). *The foundations of Wittgenstein's late philosophy.* New York: Manchester University Press.

Squires, J.E.R. (1968). Visualizing. *Mind, 77,* 58-67.

Titchener, E.B. (1910). *A textbook of psychology.* New York: MacMillan.

Titchener, E.B. (1914). *A primer of psychology.* New York: MacMillan.

Toulmin, S. (1971). The concept of "stages" in psychological development. In T. Mischel (Ed.), *Cognitive development and epistemology.* New York: Academic Press.

Vygotsky, L.S. (1962). *Thought and language.* Cambridge: The M.I.T. Press.

Wittgenstein, L. (1953). *Philosophical investigations.* Oxford: Basil Blackwell.

Yuille, J.C., & Catchpole, M.F. (1977). The role of imagery in models of cognition. *Journal of Mental Imagery, 1,* 171-180.

CHAPTER **8**

The Futility of a Purely Experimental Psychology of Cognition: Imagery as a Case Study

John C. Yuille

I. INTRODUCTION

Perhaps the title of this chapter should have been, "The Confessions of a Heretic." For almost two decades, I have enjoyed a career as an experimental psychologist. However, over the course of the past few years I have come to the conclusion that experimental psychology is generally a futile enterprise. My thesis is a straightforward one: We made a mistake when we committed ourselves to the experimental method, and it is high time that we recognized our error and set about using effective methodologies. The study of human cognition is not feasible in the laboratory context. Rather than drawing examples from the entire field of cognitive research, this assertion will be defended by reference to the experimental literature on mental imagery. Imagery continues to enjoy a central role in contemporary cognitive psychology. Not only is there a large experimental literature concerned with mental imagery, but the topic has held center stage in current epistemological debates (cf, Yuille & Marschark, 1983a).

Although mental imagery is the focus of this chapter, I must

197

emphasize that this is not another attack on imagery theories by a proponent of a propositional model. The purpose of this chapter is to provide a general indictment of the experimental method in cognitive psychology, regardless of the theoretical orientation of the experimenter. Whether used to defend an analogue or a propositional model of cognition, experiments are ultimately futile (see Yuille, 1983). The defense of this assertion is organized in the following fashion:

(1) I begin by placing my critique in the context of other criticisms of the current state of psychology.

(2) Then, the central concern of this talk is introduced with a definition of experimentation.

(3) There follows a brief discussion of the origins and development of the experimental method in psychology.

(4) The subsequent section outlines the problems inherent in the experimental method, with emphasis on the restricted nature of the knowledge generated by this method.

(5) I then turn to experiments concerned with mental imagery to provide examples of the insurmountable difficulties associated with experimentation. This section includes a review of several imagery paradigms, particularly mental rotation, mental scanning, mental comparisons, and learning and memory research.

II. DOUBTS ABOUT THE STATUS OF PSYCHOLOGY

Doubts about the status and role of psychology are not new. Since Wundt and James led the attempt to separate psychology from philosophy by encouraging the use of experiments for some psychological questions, debate has continued about the appropriate methodology for psychology. The continuing nature of this debate is a clear indication that the discipline has not developed consensus in this regard. It may also be taken as defacto evidence that no methodology has proven to be an unqualified success. Those who have chosen to be critical of the status quo have raised a variety of concerns. For example, some have questioned the applicability of laboratory findings to real life contexts. Thus, Arnold (1976), in his introduction to the text reporting the 1975 Nebraska Symposium, wrote:

To the extent...that psychology has been committed to the older, Newtonian conceptual system (and its interlocking philosophical assumptions), it becomes apparent that we, in our domain, have been guiding ourselves by a limited conception of science and, accordingly, by a restricted conception of the human being......Like physics before it, psychology is engaged in a critical reappraisal of its fundamental assumptions concerning the nature of knowledge and the very nature of the events it observes (p.vii-viii).

Neisser (1976) made a related argument when he questioned the ecological validity of cognitive research. For example, experiments may give emphasis to relatively minor variables. By isolating a facet out of its natural context, the factor may be given an importance it naturally would never have. Petrinovich (1979) has expressed this concern as follows:

The systematic experimental method separates the variables controlling behavior from the fabric in which they are embedded, and this destroys the pattern of correlations between variables as it exists in natural situations (p. 375).

We have distorted our view of ourselves to fit the requirements of the experimental method. We have created artificial environments to learn how we behave in natural environments. We have isolated variables from the context of their occurrence. The consequence is the peculiar nature of the psychology experiment. Bronfenbrenner (1977) provided an excellent description in the context of an evaluation of developmental research:

(T)he emphasis on rigor has led to experiments that are elegantly designed but often limited in scope...much of contemporary developmental psychology is *the science of the strange behavior of children in strange situations with strange adults for the briefest possible periods of time* (p. 513).

In most cognitive experiments, the object of study is adults; but the context is just as strange as Bronfenbrenner's children encounter. The current critique accepts that the artificiality of the laboratory is a major problem for experimental psychology. I wish to add another concern to the list. Because of the dynamics of the experiment, together with some of our publishing practices, it is possible to prove, experimentally, any reasonable hypothesis about human cognition. In skilled hands, the experiment is not a means of testing hypotheses, it is a way of proving them:

...what the experimenter tests is not whether the hypothesis is true but rather whether the experimenter is a sufficiently ingenious stage manager to produce in the laboratory conditions which demonstrate that an obviously true hypothesis is correct (McGuire, 1973, p.449).

Before elaborating this fundamental flaw in the experimental method, it is appropriate to provide a definition of that method.

III. THE EXPERIMENTAL METHOD

I begin this definition with a qualification. My anti-experimental thesis is not grounded in anti-empiricism. In fact, my thesis has developed from a firm belief that psychology can be and should be a science. It is with one particular philosophy of science that I find fault: positivism and the experimental methodology associated with it. In contrast, research in the form of systematic observation *in situ* is valid and useful. Such research can improve our knowledge and understanding of mental processes. But I am getting ahead of myself, and I will reserve an elaboration of appropriate methodologies for the end of this chapter. My present concern is with the definition of the experimental method I wish to criticize.

By the term *experiment*, I am referring explicitly to laboratory-based hypothesis-testing in which the purpose of the research is the clearest possible demonstration of causality. This is the methodology which is taught in most of our research design courses and texts, and which is outlined in our introductory psychology books as the appropriate method for the discipline. It is defined, in its simplest form, as being dependent upon the statement of a hypothesis about the presumed relationship between an independent variable and a dependent variable. The principal preoccupation of the experimenter is the exercise of control. In what fashion can all aspects of the experimental setting be controlled so that any variation in the observed behavior of the *subjects* of the research can be unequivocally attributed to variation in the independent variable? If control is adequately exerted, we will learn the contribution of the manipulated variable to changes in behavior. Of course, the latter attribution is aided by an impressive battery of inferential statistical procedures, particularly analysis of variance. Once a causal relationship is demonstrated, some mediating cognitive process can be inferred. In fact, the inference about intervening cognitive states or processes is the ultimate purpose of the experiment (Yuille, 1983). The variables chosen for attention in the experiment are presumed to be related or to affect our cognitions.

The experimental method is a procedure which we adopted in the nineteenth century from physics and chemistry. The combination of control and systematic variation of variables had proved appropriate for the basic sciences, and these disciplines stood as obvious role models for the first generation of psychologists. As these academics attempted to establish psychology as a separate discipline, many were charmed by the success of the experimental method in providing knowledge of the physical world. Considering the compatibility of the method with the prevailing empiricist epistemology, it seemed obvious to many that our own nature would be equally yielding to the probes of experimentation.

During the first decades of psychology's independent existence (roughly 1880-1920), the popularity of the experimental method was widespread but it was not universal. Several of the founding fathers of psychology, for example, Wundt, James, and Brentano, saw a limited use for experiments in the new discipline. For example, Wundt thought that experimentation was useful only in the study of the current contents of consciousness, which he called the outer aspects of mind. These included perception, attention, and memory. Wundt argued that the more complex aspects of cognition, for example, problem solving, language, etc. could not be studied experimentally. Instead, he proposed that psychologists would need to examine the behavior of people in their cultural context in order to infer the nature of complex cognitions. He called this branch of psychology *Volkpsychologie*. For example, he provided the following definition of the methods of psychology:

Thus, psychology has...*two* exact methods: the experimental method, serving for the analysis of simpler psychical processes, and the observation of general mental products, serving for the investigation of the higher psychical processes and developments (Wundt, 1902, p. 27-28).

To a limited extent, the argument presented in this talk is similar to the one Wundt articulated a century ago. Experimentation may be of use to us in the investigation of simple psychological processes, but it is wholly inappropriate for the study of complex processes. Wundt believed that experiments applied to some aspects of psychology because of his belief in the central role of psychology to all of science. He argued that since physics and chemistry depended upon reported human observation, that the same method should be appropriate for reporting the contents of consciousness. My argument is different. It is based upon the assumption that most aspects of cognition are context-dependent, and that the laboratory provides a peculiar, not an appropriate, context for the study of cognition. The only psychological processes which we can comfortably study via the experiment are those which do not change by virtue of their being studied in the context of the laboratory.

In the early part of this century only a few psychologists, particularly in North America, were doubting the appropriateness of the experimental context. Instead, the promise of objective knowledge combined with the progressive movement in America to produce an optimistic hope for an experimentally based psychology. Careful, rigorous experimentation would reveal the laws which controlled human behavior, and we would become the masters of our own destiny. In the pursuit of this hope, the experimental method was adopted, adapted, and developed. Positivism, labeled as operationalism, became the official philosophy of the new science. The experimental methodology, the associated technology and statistical techniques, all became increasingly sophisticated. But the hope of behaviorism remained unfulfilled. The search for behavioral laws proved futile. Even the impressive theoretical attempts of neo-behaviorists like Tolman and Hull failed.

The failure of behaviorism provided an opportunity for psychology to realize the limitations of the method and philosophy it had borrowed from nineteenth century basic science. But the opportunity was missed. Perhaps this reflects what someone has called the methodolatry of psychologists. The method was not questioned, only the behaviorist exclusion of mentalistic terminology from psychology. Beginning in the 1950s and accelerating through the 1960s, cognitive terms reappeared in the discipline. Images, ideas, plans, feelings, were once again part of the psychologist's lexicon. Many of us shared a new hope: that the established, sophisticated, experimental method together with the new, cognitive vocabulary would provide the psychological insights which had proved so elusive.

This hope remains unfulfilled, and the time is long past for an admission that it cannot be fulfilled. The experimental method is wrong for the study of cognition primarily because cognitive processes are con-text-dependent. That is, we are malleable organisms, capable of adapting ourselves to the perceived requirements of a situation. The experimental situation is a contrived context in which the experimenter has complete control of the information available to the *subject*. The experimenter has designed the experiment to test, which usually means support, an hypothesis. Everything about the situation is construed so that the most likely behavioral outcome of the experiment is the one which will support the hypothesis. The subject, who is a cooperative volunteer, is actively trying to determine how to behave in this novel context. The instructions provided, the materials employed, the environment used, the measures recorded, etc. combine to limit the behavioral options of the *subject*. As a consequence, the results of most published experiments confirm the experimental hypothesis. But we are deceiving ourselves when we believe that these results are providing insights into cognition. Rather our experiments are demonstrating the willing malleability of *subjects*, and the cleverness of exprimenters in constraining responses. William James (1890) stated that "the mind is at every stage a theater of simultaneous possibilities" (p.24). The laboratory provides a demonstration on how to effect different possibilities. Furthermore, the laboratory realizes those possibilities which are unique to the laboratory. Experimental results tell us nothing about ordinary (or even non-laboratory extraordinary) cognition; they show us the flexibility of human cognition and behavior. Nothing more.

A colleague, Ray Corteen, has provided an excellent metaphor for the role of the experiment in cognitive psychology. The experimental psychologist can be compared to an investigator of human mobility who is convinced that people move about in their environment by crawling. He builds a constraining environment within which to observe human movement. The experiment is carried out in a room with a one-foot ceiling. The only movement possible for the *subject* is wriggling about in a prone position. The experimenter's hypothesis is confirmed, and he continues with the belief that because of the objectivity of the experimental method, he has contributed to human knowledge. Yet no useful knowledge has been generated in this parody of an experiment. In an analogous fashion, we constrain cognitive and behavioral options in our laboratories, and the problem is that the constraining nature of the laboratory is intrinsic. No modification in experimental procedures can alter this fact.

An experiment is an elegant, sophisticated method of controlling human behavior. It is a procedure by which we can mold, shape, and manipulate individuals to maximize the likelihood that an experimental hypothesis is supported (e.g. Bannister, 1977). Consider the typical way in which we do experiments. We generate an hypothesis, for example, that mental images share certain qualities with perception, specifically, that imaged scenes require time to scan, just as perceived scenes do. We select a set of materials to give *subjects*, we prepare a set of instructions, determine a laboratory context, and decide on the dependent measures. We try the experiment, but the results which we are obtaining are not what we hoped; we are not finding support for the experimental hypothesis. So we call this a "pilot study," change the materials, or the instructions, and repeat the experiment. This process is continued until we have succeeded in the "right" combination of experimental factors to get the desired result. This final study becomes the experiment which we report, and the previous experiments are ignored. Consider the following quote from a recent paper by Finke & Pinker (1983) which was concerned with demonstrating the time required to scan mental images: "...we noted that several of our early pilot experiments had failed to show evidence for mental image scanning....We were therefore interested to see whether we could influence the selection of strategies" (p. 403). As is true for most published experiments, these authors were successful in finding the right combination of factors to produce the behavior they needed to support their hypothesis.

This last remark may leave the impression that I am attributing an experimental machiavellianism to cognitive psychologists. This is not the case. Clearly, most experimentalists are operating with the sincere belief that useful, objective knowledge can be generated in the laboratory. Successive iterations of an experiment until the desired outcome is obtained is considered legitimate science. My argument is that this belief is misguided. Because of the malleability of human cognition and the dynamics of experimental control, a patient experimenter will find support for his/her model, not because of experimenter bias, although this may play a role, but because this is the only possible outcome of psychological experiments. Our hope that what people do in artificial tasks in an artificial context will tell us something about our cognitive nature will remain unfilled. It is appropriate to explore some specific examples at this point to seek support for these assertions.

IV. MENTAL IMAGERY EXPERIMENTS

In this section I review several of the experimental paradigms which have been used during the past two decades to investigate the nature of mental imagery. The paradigms are mental rotation, mental scanning, mental comparisons, and verbal learning. I've chosen these because they are representative of research done in this area, and because they reflect different problems associated with experiments in cognitive psychology. The four paradigms demonstrate the common problem of experimentation, but each paradigm demonstrates a particular form or manifestation of the general problem.

1. Mental Rotation

No doubt, the most cited work in the imagery area in the last decade is that of Shepard and his colleagues on mental rotation. This paradigm originated with a publication by Shepard and Metzler in 1971. Using pairs of figures like those shown in Figure 1, they asked *subjects* to indicate whether the two figures were the same, differing only in angular rotation (as in the bottom pair in Figure 1), or whether the two figures were different. The different pairs were mirror images of one another, as well as differing in angular rotation (the top pair in Figure 1). The time required to correctly identify two figures as the same was found to be a linear function of the angular disparity of the two figures. The typical finding is portrayed

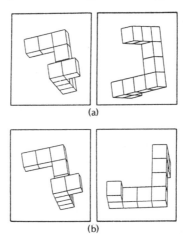

(a)

(b)

Figure 1

in Figure 2. Note the strong linear relationship betweem reaction time and angular disparity of pairs of figures. Shepard (e.g., 1978) has interpreted this linear function as demonstrative of the role of mental imagery in this task. He argues that images require time to rotate, and that this time is

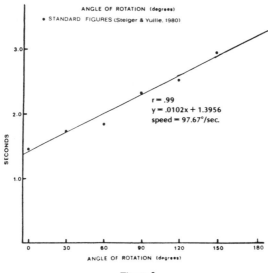

Figure 2

reflected in the slope of the regression line. Indeed, Shepard has estimated the speed of mental rotation from the slope of his reaction-time functions. Shepard believes that these, and related results, provide evidence for the quasi-perceptual nature of mental imagery. The image, he suggests, represents such properties as distance, size, and motion in ananalogue fashion. Unfortunately, Shepard's results are due to the materials and procedure he employed. Only minor changes in the stimuli produce different results. As a consequence, whether mental imagery is involved in this task is indeterminate. Similarly indeterminate, from the results of mental rotation experiments, is whether images have the properties Shepard has attributed to them.

The stimulus-dependent nature of Shepard's experimental results was revealed in the following fashion. Shepard argued that images are holistically rotated, so that the complexity of the image is irrelevant to the speed of mental rotation. A colleague, Jim Steiger, and I noticed a figural redundancy in one set of the block figures Shepard had employed (see Figure 1). For these figures, it is necessary to examine only the bottom arm of each figure in order to make a comparison. This is illustrated in Figure

3, which displays only the bottom arms of the figures previously shown in Figure 1. Shepard had mixed different figures together so that his *subjects* would not have discovered the redundancy. We wondered what the consequences of using these redundant figures would be. We (Yuille & Steiger, 1982) used a homogeneous figure set, and informed half of our *subjects* of the redundancy. On the basis of Shepard's holistic rotation hypothesis, this information should not have affected the rate of responding in this task, but it did. As the next two figures show, informing a person of the need to examine only part of the figure reduced the time required to perform the task. Figure 4 shows the data for the *subjects* who

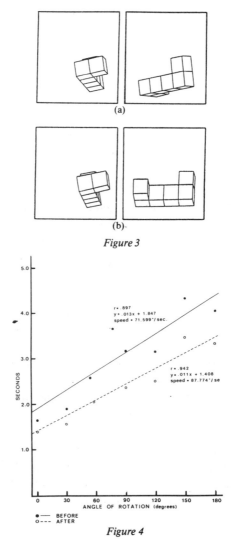

(a)

(b)

Figure 3

Figure 4

were not informed of the redundancy in the figures. The two curves are for the first half of the trials and the second half of the trials. The difference represents practice effects. Figure 5 shows the comparable data for the

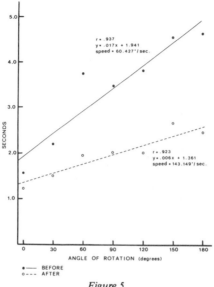

Figure 5

experimental group. These *subjects* were informed after completing the first half of the experiment of the redundancy. This information resulted in a much larger decrease in response raters than the practice effects in the control group. That is, informing people halfway through an experiment of the redundant properties of this type of block figure allows them to compare the figures more rapidly.

Not only does simplifying the figures change behavior, making the figures more complex results in substantial behavioral changes as well. We (Yuille & Steiger, 1982) created more complex block figures like those shown in Figure 6. As you can see, the same redundancy which Steiger and I exploited in our previous research is built into these larger block figures. When shown pairs of such figures, many of our *subjects* couldn't do the task. For these people, changes in complexity had a profound effect: the task became impossible. For those who were able to deal with the more complex figures, they reported that they could do so because they spontaneously discovered the built-in redundancy. They made the task manageable by making the complex figures into simple ones.

To determine what would happen if such simplification were made impossible, we developed some figures like those in Figure 7 in which one segment of the figure was twisted. These figures do not permit a

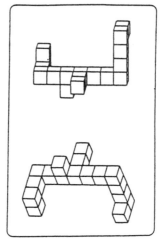

Figure 6

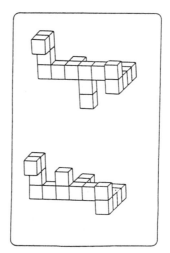

Figure 7

simplification strategy. As the regression lines in Figure 8 indicate, reaction times were different for the figures which contained a twisted segment. If rotated mental images mediated this task, a twisted segment in a complex figure should not affect reaction time. Mental rotation, in the sense that Shepard has used the term, is not the mediating process in the "mental rotation" task.

The point of these experiments is clear. The size of the figure, and

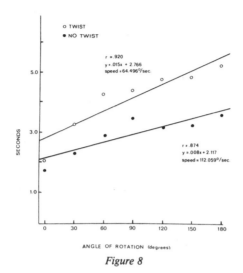

Figure 8

the use of only mirror images as different figures produced the regression line function in the mental rotation paradigm. If Shepard had employed a simpler figure, the conclusion that mental rotation takes time would not have been made. If he had employed more complex figures, his conclusion that mental images are holistically rotated could not have been made (see also Pylyshyn, 1979). If Shepard had included twisted segments in his different figure pairs, he could not have concluded that imagery was operative at all in this task. The right combination of stimuli and instructions produced the hypothesis-confirming results. In fact, all we know from this "mental rotation" pardigm is that there is a linear relationship between correct reaction time and angular disparity for particular figures in a particular task (see also Steiger & Yuille, 1983). We know nothing about the mediating processes. Even if we did it is clear that the finding has no generality, even to related figures in a slightly different task. The truth is that we have learned nothing about mental imagery from so-called mental rotation experiments.

2. Mental Comparisons

The imagery paradigm with the greatest amount of face validity is the mental comparison task. An individual is asked to make some comparison between two objects, based upon his/her memory for those objects. The basis of the comparison is some dimension which has a visual quality. The purpose of this paradigm is to find percept-like qualities in memory so that analogue characteristics, that is, images, can be attributed to memory. Mental comparisons have taken a variety of forms. The paradigm was introduced by Moyer in 1973. His *subjects* were presented with pairs of animal names, and were required to decide which member of the pair is larger in reality. Moyer found an inverse relationship between speed of answer and the actual size difference between the animals. We know from psychophysical research that perceptual comparisons require more time when objects are more similar in size. Moyer's demonstration of a similar effect for mental comparisons was taken as evidence for the existence of mental images. He suggested that an analogue comparison must mediate the task, and that it is more difficult to compare images which are similar in size; hence, longer reaction times.

The next few figures provide demonstrations of this paradigm and typical research findings. They have been adopted from reports of Paivio (1975; 1978a). He compared *subjects'* capacity to make size comparisons from pictures and from words. Figure 9 shows examples of the type of pictures he used. One pair of pictures presents the figures in proportions

Figure 9. Examples of congruent (A) and incongruent (B) size relations between pictured items in the physical size - memory size conflict study.

Note. From "Images, propositions, and knowledge" by A. Paivio, *The Western Ontario Series in Philosophy of Science*.

which are congruous with their relative sizes. In the other pair, the relative sizes are incongruous with reality. A similar manipulation was made by Paivio with verbal stimuli, as Figure 10 illustrates. The findings from this experiment are shown in Figure 11. Congruous pictures were the easiest to deal with, and incongruous pictures caused a delay in responding. Congruity had no effect on size comparisons made from word stimuli, but

A

LAMP   **ZEBRA**

B

LAMP   ZEBRA

Figure 10. Examples of congruent (A) and incongruent (B) size relations in the word condition of the physical size - memory size conflict study.

Note. "Images, propositions, and knowledge" by A. Paivio, *The Western Ontario Series of Philosophy of Science.*

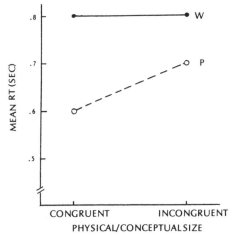

Figure 11. Reaction time for choosing the conceptually larger of two objects presented as pictures (P) or as words (W) when physical size and conceptual (memory) size differences are congruent and when they are incongruent.

Note. From "Images, propositions, and knowledge" by A. Paivio, *The Western Ontario Series in Philosophy of Science.*

word pairs required longer time to make the size decisions. Paivio argued that the words must be translated into images before the size comparison can be made. Hence, the words require more time.

Other mental comparison tasks have produced similar results. For example, Paivio (1978a) required *subjects* to make judgments about the angular separation of the hands of clocks. Pairs of clocks were compared, and the *subject* had to indicate which pair of hands had the smallest separation. In some cases, labeled "analog" in Figure 12, the hands were shown. In other instances, only the digital values were given, as shown at the top of the slide. Finally, some comparisons involved a mixture of the analogue and digital forms. Angles of similar size, for example, 3:20 and 7:25, took longer to discriminate than strikingly different angles, for

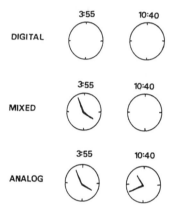

EXAMPLES OF STIMULI USED IN THE THREE CONDITIONS

Figure 12

Note. From "Comparisons of mental clocks" by A. Paivio, 1978, *Journal of Experimental Psychology: Human Perception and Performance.*

example. 4:10 and 9:23. Figure 13 displays the results of one of these experiments. Note a slight decrease in latency as the angular difference between the times increased. The effect is slight but reliable. More important is the much longer time for digital as compared to analogue comparisons, and the intermediate times of the mixed condition. It was

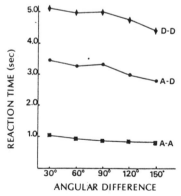

Figure 13. Mean reaction time as a function of angular size differences for digital-digital (D-D), Analog-Digital (A-D), and Analog-Analog (A-A) comparisons.

Note. From "Comparisons of mental clocks" by A. Paivio, 1978, *Journal of Experimental Psychology: Human Perception and Performance.*

argued that the digital format takes longer because it must first be translated into analogue form before a comparison can be made. Again, the quasi-perceptual nature of images was invoked to explain the results.

In a similar vein, Shepard and Chipman (1970) had volunteers rank order the American states in terms of the similarity of their shapes. The first ranking was done only with the names of the states, and the second with the pictures of the states. There was a strong correspondence between the two rankings. Shepard and Chipman felt that this must be because analogue images were used to make the comparisons for the first, verbal ranking.

The mental comparison tasks seem to provide clear evidence of an analogue capacity of memory. Given the pattern of responding it must be the case that analogue images mediate the task. However, Friedman (1978) found that it was possible to produce the same pattern of results when an abstract dimension is employed. He presented individuals with the word pairs generated on the basis of semantic differential ratings on the good/bad dimension. The task for the experimental participants was to decide which member of the pair was either "better" or "worse" than the other member of the pair. Examples of the pairs are "hate-peace" and "argue-freedom." The closer the two words of a pair were on the "good/bad" scale, the longer it took people to differentiate the words. Thus, a pattern of responding which was interpreted as reflecting the mediating role of mental images was found for abstract properties of words. Friedman (1978) felt obliged to conclude that "it is entirely possible for memorial comparisons to yield these (linear) functions without recruiting the mind's eye" (p. 443). The linear function relating size and reaction time is not a demonstration of mental imagery, it demonstrates that people have more difficulty discriminating closely related stimuli than disparate stimuli. By presenting the right type of stimuli and instructions, we have been tempted to attribute the discrimination difficulty to the analogue nature of memory. But this is unwarranted, and can lead to circular argumentation. An example is found in Paivio's (1978b) suggestion that abstract qualities such as pleasantness must be represented imaginally, since they produce symbolic distance effects. Whether imagery is employed at all in these tasks was made doubtful by an experiment by Holyoak (1978), who found that explicitly instructing *subjects* to use imagery when making mental size comparisons resulted in longer reaction times than uninstructed individuals.

All we know from this research is that mental comparisons take time, and that the more similar two items are on a dimension, the more difficult it is to discriminate them on that dimension. We have also learned that there are a number of experimental procedures which influence performance in a mental comparison task (see Marschark, 1983). But we have learned nothing about mental imagery.

3. Mental Scanning

Kosslyn and his associates (e.g., Kosslyn, Ball & Reiser, 1978) have introduced a paradigm which is related to the mental rotation and mental comparison tasks in that the *subject* is asked to form an image, and reaction times are used to infer something about the nature of the image. In one version of this technique, individuals are shown stimuli such as those in Figure 14. They are asked to focus on their mental image of one of

EXAMPLES OF LINE DRAWINGS USED

Figure 14

Note. From "Scanning visual images: Some structural implications" by S.M. Kosslyn, 1973, *Perception & Psychophysics, 14,* 90-94.

these figures, and scan the image from one point to another. The time required to report the completion of such a scan is found to be a linear function of the distance scanned (see Figure 15). In another version of this paradigm, the image is of a map or an array, and the task is to mentally focus upon one part of the map, and then move to some other part. For example, a person might be asked to memorize a map like that shown in Figure 16. Subsequent to memorization, he/she will be asked to focus on one landmark, like the lake, and then to move an imagined point from the lake to the hut, and to press a button when the mental journey is completed. In a series of experiments, Kosslyn and others have reported that reaction time is directly related to the distance between origin and destination (see Figure 17). Once again, reaction-time changes are interpreted as indicative of the analogue nature of mental images. Images, like percepts, require real time to scan.

This paradigm is ideal for the critic of the experimental

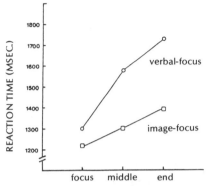

Figure 15. Speed of verification of picture properties as a function of distance from the point of focus.

Note. From "Scanning visual images: Some structural implications" by S.M. Kosslyn, 1973, *Perception & Psychophysics, 14,* 90-94.

THE FICTIONAL MAP

Figure 16

Note. From "Visual images preserve metric spatial information: Evidence from studies of image scanning" by S.M. Kosslyn, T.M. Ball, and B.J. Reiser, 1978, *Journal of Experimental Psychology: Human Perception and Performance, 4,* 47-60.

methodology. One issue that is revealed here concerns the selective inclusion of data in the final report. Due to the peculiar nature of the task demands, that is, scanning mental images, some people find the experimental demands difficult to meet. Consequently, a proportion of the *subjects* cannot perform the task to the experimenter's desired criteria. These people are simply discarded from the experiment, and only the data from the more cooperative *subjects* is retained. In one of Kosslyn's studies, 25% of the people he tested were rejected for not following instructions. This is an excellent example of what I noted earlier about the controlled

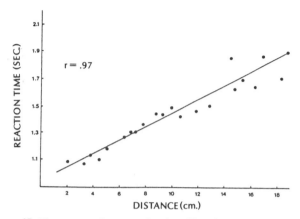

Figure 17. Time to scan between all pairs of locations on the imaged map.

Note. From "Visual images preserve metric spatial information: Evidence from studies of image scanning" by S.M. Kosslyn, T.M. Ball, and B.J. Reiser, 1978, *Journal of Experimental Psychology: Human Perception and Performance, 4(1), 47-60.*

nature of the experimental setting. Since everyone is not going to behave in the fashion the experimenter desires to demonstrate, only those who do behave in the desired way will be considered.

There is an additional experimental difficulty revealed by mental scanning studies. This paradigm readily displays the role of *subjects'* knowledge in affecting the outcome of the experiment. Charlie Richman at Wake Forest has shown that he can get the same results as Kosslyn and his associates without doing the experiment (Richman, Mitchell, & Reznick, 1979; Mitchell & Richman, 1980). Richman provided volunteers with a description of the mental scanning task, and asked them to estimate how people would perform. This task is what Pylyshyn (1980) would call cognitively penetrable, and Richman's pseudo-subjects were able to correctly anticipate how Kosslyn's *subjects* performed. It is so apparent from the nature of the task, the instructions, and the dependent measure, that the experimenter is expecting different latencies for different distances, that the cooperative *subject* can readily provide the desired behavior.

The problems of rejecting uncooperative *subjects* and task demands are sufficient to question the value of this type of research. These problems have been amplified by the explicit demonstration of experimenter bias effects in mental scanning research. The role of experimenter expectancies in mental scanning has been reported in two papers by Intons-Peterson. The first of these (Intons-Peterson & White, 1981) concerned a finding of Finke and Kurtzman (1981) that both imagined and perceived stimuli were affected by changes in size while a change in brightness only affected a perceived stimulus. They argued that size information is property that images and percepts share, while brightness is a more sensory quality which only affects percepts. Intons-Peterson and White employed an experimenter who was naive concerning any expectations about the paradigm. Naive experimenters did not replicate Finke and Kurtzman, with imagined stimuli being unaffected by both brightness and size changes. Subsequently, Intons-Peterson (1983) varied the expectancies of naive experimenters in different imagery paradigms (mental scanning and mental rotation). In each case, expectancies had a strong effect, even when contact between the experimenter and the *subject* was restricted. She concluded that pauses in speech, inflections, and animation differences helped to control the behavior of the *subjects*.

Intons-Peterson (1983) concluded that "we must devise paradigms that exclude as many alternative explanations to imagery as possible" (p. 411). My suggestion is that such a paradigm is impossible to devise. It is intrinsic to the experimental method that the subject is under the control of the experimenter and the setting the experimenter has devised. Nothing can change this except to leave the laboratory, and begin to observe human behavior *in situ*. Only minimum intervention on our part will assure that we are not creating the behavior we are observing.

5. Verbal Learning Experiments

The renewed interest in mental imagery experiments in the 1960s was reflected primarily in verbal learning research. This was due to the fact that paired-associate learning, free recall, and serial learning were the popular paradigms for researchers before imagery became a topic of interest. In order to demonstrate the necessity of including imagery in the theoretical lexicon of psychology, researchers were obliged to demonstrate its role in verbal learning. The work of Alan Paivio was particularly valuable in this regard. First, he and his colleagues provided an operational definition of imagery. This was achieved by having groups of people rate the capacity of words to elicit mental images. The ratings of 925 words provided by Paivio, Yuille, and Madigan were published in 1968, and they are still widely used. Paivio (1971) demonstrated that rated imagery is a potent variable in paired-associate learning. In addition, Paivio (1971) has demonstrated that when individuals are asked how they learned paired-associates, those learned via imagery are more likely to be recalled than those reported learned by rote repetition. The combinatory effects of rated imagery, imagery instructions, and reported use of imagery provide what Paivio (1983) believes is converging evidence of an important role of imagery in human memory. He has proposed his dual coding hypothesis to explain this stimuli. He has hypothesized two systems, an imagery system specialized for coding parallel relations in analogue form, and a verbal system specialized for coding serial information in a linguistic form.

There are two major difficulties associated with the experimental work on imagery in human memory. The first of these is the validity of the inference that imagery can mediate memory. The convergent evidence that Paivio (e.g., 1983) and other imagery theorists rely upon appears to have face validity. If people are asked to rate the image evoking capacity of nouns, and these ratings, in turn, are predictive of ease of learning of those nouns, it seems appropriate to attribute the memory differences to imagery. Recently, Zimler and Keenan (1983) reported three experiments in which the memory of congenitally blind children and adults was compared with that of sighted children and adults. In the first of these experiments, the *subjects* were asked to learn pairs of nouns which varied in their ability to evoke visual images. Zimler and Keenan assumed that the blind would show a deficit in their ability to code visual nouns, as Paivio and Okovita (1971) had previously reported. On the basis of their results as well as other reports, Zimler and Keenan concluded that "there appears to be no consistent evidence for an impairment in the blind's memory for words with visual referents either relative to the sighted or relative to other word types" (p. 273). Obviously, the face validity of the imagery ratings of nouns faces a serious challenge with this conclusion.

Another aspect of the converging evidence implicating imagery in human memory concerns the effect of imagery instructions. However, Jonides, Kahn and Rozin (1975) found that imagery instructions were as effective in improving memory for congenitally blind *subjects* as for sighted *subjects*. More recently, Kerr (1983) reported three experiments with congenitally blind adults which placed the face validity of imagery instructions in doubt. She found that the pattern of performance of sighted and blind *subjects* was the same in almost all respects. She concluded that "the comparability of the data from blind and sighted subjects provides convincing evidence that...performance in these imagery tasks (is) not uniquely dependent on the visual processing system" (p. 273).

The results of experiments with congenitally blind individuals raises doubts about the imagery interpretations of memory experiments. Another example of the equivocal nature of such interpretations is found in the research reported by proponents of propositional or computational models of memory (cf., Yuille and Marschark, 1983b). For example, Anderson (1976) has suggested that both imagery instructions and the rated imagery of words are related to memory because of their affect upon meaningfulness. That is, imagery instructions encourage us to elaborate more meaningful relations between unrelated words. Thus, both pro - and anti-imagery theorists find support with the same data base. Paivio's (1983) convergent evidence argument has some force, but it rests upon shaky ground.

In summary, it is not clear that mental images mediate human memory in laboratory situations. But even if one ignores this difficulty, there is an additional problem associated with memory experiments. This difficulty reflects the questionable ecological validity of the research results. When designing paired-associate, or similar, laboratory-based learning tasks, the experimenter makes every effort to control the meaning of the learning materials. While this meets the control requirements of the experimental method, it raises the possibility that the resulting material is not representative of what human beings typically learn. Impoverished stimuli, whether a pair of words or a set of sentences, do not reflect the richness or complexity of much of what we learn. In order to learn laboratory materials, we must provide a type of elaboration that may be unique to the laboratory. Thus, even if one accepts a role for imagery in the coding of verbal materials, how much this tells us about the normal operation of human memory is indeterminate. Let us assume that imagery is an effective technique in mediating the learning of impoverished materials; what does this tell us about the normal operation of human memory? Our problem is that there has been little attention paid to this question. In spite of the fact that as teachers we are constantly engaged in the learning process with students, we don't test the efficacy of our ideas in the complex context of the classroom. Would imagery instructions improve the learning of psychology by undergraduate students? The strange thing is that we don't know. Paivio (1982) has indicated that when he was learning French, imagery was only useful during the initial stages of vocabulary acquisition. This observation suggests a much more limited role for imagery in memory than the verbal learning research would indicate. I have one finding to report that bears upon this.

A graduate student under my supervision, Hope McEwan, did her Master's thesis research on the value of hypnosis as an aid to eyewitness memory. Hypno-investigators in the field typically use imagery instructions as part of their technique for guiding witnesses during hypnosis. Thus, as part of this investigation, some of our participants received imagery instructions (one-third of these were also hypnotized, the rest were not). Imagery instructions had no effect on eyewitness recall or recognition. This is only one study, and the inclusion of imagery instructions was tangential to the purpose of the research; however, since I am used to strong imagery effects in verbal learning studies, it is striking to find no effect when more complex aspects of memory are tapped. Until we complete the research necessary to ascertain the role of imagery in ecologically valid contexts, we must maintain the possibility that the domain of imagery is the laboratory.

V. CONCLUSIONS

This brief examination of experimental paradigms ostensibly used to investigate imagery reveals a bleak picture. Due to the controlled nature of the experimental setting, the artificial attempt to isolate some aspect of cognition and study it, and the malleability of human cognition, the experiment becomes a tool of hypothesis confirmation, and not a method through which we can investigate psychological processes. The mental rotation research has revealed how sensitive we are to changes in the characteristics of the stimuli we employ. The right choice of stimuli will produce behavior consonant with a particular theoretical position. The mental comparison research shows the ease with which unjustified conclusions can be drawn from the apparent face validity of the task. The mental scanning tasks demonstrate that the structure of the experiment - in particular, the implicit demands that a task makes of a *subject* - can produce the behavior the experimenter is seeking. As soon as a researcher intervenes in human affairs, the risk of these task demands is present. The greater the degree of intervention, the greater the risk. Our mistake is that we are seeking precision, a goal which is wholly inappropriate for psychology. The more precise our findings, the more irrelevant they are. The goal of precision was appropriate to nineteenth century physics; the phenomena they were concerned with were not changed by virtue of their study in the laboratory. The physicists had not yet reached the point where the position of the observer influenced and limited what could be observed. The relativity of psychological knowledge is more pervasive. As soon as we try to isolate some phenomenon in the laboratory, we change it. Unless we can be sure that this is not the case for any particular phenomenon, no experiment should be considered. Experiments should be the exception rather than the rule in our research. The litmus test for the appropriateness of the laboratory context should be the effect of that context on the behavior of interest. The experimenter should be obliged to demonstrate that the phenomenon of interest is not changed by virtue of bringing it into the laboratory. Since we must accept, as far as cognition is concerned, that this will rarely be the case, experiments should rarely occur. Some aspects of perception, particularly those that used to be labeled sensation, and some psychophysical investigations may be laboratory proof. But, as

Wundt argued a century ago, none of the higher cognitive processes are amenable to experimental investigation. We must examine the products of cognition *in situ* if we hope to learn anything about the nature of mind. The appropriate balance of laboratory and non-laboratory research will be empirically determined. The essential point is that the boundary conditions of psychological knowledge about cognition will be established by *in situ* research.

As the quotations included in my text have indicated, dissatisfaction with experimental psychology has been expressed by a variety of people. Although I have defended my thesis with examples of mental imagery research, I could have chosen other aspects of cognition, as Neisser (e.g., 1976) has done. Also, my criticisms could have been based on developmental research (e.g., Bronfenbrenner, 1977) or social psychology experiments (e.g., McGuire, 1973). The shortcomings of the experimental method are not confined to studies of imagery. Alternatively, I could have cited the criticisms of Deese (1969) or Jenkins (1980) or Koch (1981) or Stent (1975). Criticisms like those I have outlined have been articulted by others. My interpretation of the literature is that the number of such expressions is growing. Informal discussions that I have had at conferences has reinforced my view that many believe that a change is necessary. Currently, the editors of the journals of the American Psychological Association are discussing ways of discouraging the publication of the typical experiment. They are seeking a way to have research reports "tell a story." Another symptom of change is the recent debate in social psychology concerning whether the subdiscipline should be viewed as science or history. Similarly, developmental psychologists are examining dialectics and hermeneutics as alternative methodologies to positivistic science. A recent article by Ahsen (pages 1-45 of this book) examines the historical roots of this methodological issue in psychology concerning the notion of experience, and makes a plea for an empirical departure which puts belief system and myths squarely within the pure view of empirical psychology. His reevaluation of the historic implications implicit within the empirical method clearly enriches our traditional concept of developmental methodologies, even social psychology, history and politics. A change is under way in psychology, although its existence and direction have not been clearly articulated. Certainly, if my anti-experimental thesis is correct, our major task is to determine the form that psychological studies will take in the future. I conclude with a few brief suggestions.

The change that is required in our methodology must reflect the change we must make in our theoretical aspirations. The theoretical goals of physics are inappropriate for our discipline. To quote from a recent article by Broughton (1981):

> Experimentation, in the traditional sense could never reveal the nature of psychological processes, since they are not of the logical kind that could be dealt with exclusively through the controlled manipulation of variable and the testing of hypotheses. . . .The form and meaning of thought as well as its process of unfolding are not reducible to causal mechanical relations (p. 217).

If we accept the goals of explanation and understanding, our method will not be founded upon the unequivocal demonstration of Newtonian causality. Instead, we can accept the messy but more meaningful data that people provide every day in life. Our method ought to be that of systematic observation *in situ*. The recent text by Neisser (1982), titled *Memory Observed*, provides a number of examples of non-laboratory studies of memory. The case studies, diaries, and field studies reveal that we can learn about human memory *in situ*. They also reveal that our laboratory efforts have been little help in aiding our understanding of natural memory. I can offer a personal example. During the past few years, reflecting the concerns which I have outlined above, my own research changed from verbal learning experiments to the more complex paradigm of eyewitness research. Like most researchers interested in eyewitness performance, I began by having people witness events presented on slides or video tape. These witnesses were then asked a series of systematic questions. It didn't take too long before I realized that this eyewitness research was simply a more complex form of a verbal learning experiment, and, as a consequence, just as vacuous. More recently, my assistants and I have begun *in situ* eyewitness research. We are accompanying police on patrol, and detectives during crime investigations. We are finding that the behavior of actual eyewitnesses is different than the laboratory research had suggested. For example, real witnesses are generally better than laboratory witnesses. But, more important, the questions that are raised in the real-life context are different from those which the laboratory provided. For example, the effect of alcohol and other drugs, something ignored by laboratory investigators of eyewitness memory, is of critical concern when examining real witnesses. Before this example takes me too far afield from my present thesis, I'll simply note that observing people in a context where we, the observers, do not determine what those people do, is the only context in which we are going to learn about ourselves.

To reiterate my conclusion. The role of imagery in cognition has not been revealed in our experiments, nor can it ever be. It will be a major move for academic psychologists to leave the laboratory, but the choice is a simple one: abandon the laboratory or remain irrelevant. I think that the move has begun. There is an enormous experimental infrastructure which will resist the move, and progress will be difficult. Nonetheless, the task now before us is to acknowledge our past errors, and begin the difficult task of generating a useful data base for our discipline.

REFERENCES

Ahsen, A. (1985). Image psychology and the empirical method. *Journal of Mental Imagery, 9*(2), 1-40.

Anderson, J.R. (1976). *Language, memory and thought.* Hillsdale, NJ: Lawrence Erlbaum.

Arnold, A. (Ed.), (1976). *Nebraska symposium on motivation, 1975.* Lincoln, NE: University of Nebraska Press.

Bannister, D. (1977). On the absurdity of "being a psychologist." *Bulletin of the British Psychological Society, 30,* 211.

Bronfenbrenner, U. (1977). Toward an experimental ecology of human development. *American Psychologist, 32,* 513-531.

Broughton, J.M. (1981). Piaget's structural developmental psychology: Logic and psychology. *Human Behavior, 24,* 195-224.

Deese, J. (1969). Behavior and fact. *American Psychologist, 24,* 515-522.

Finke, R.A., & Kurtzman, H.S. (1981). Area and contrast effects upon perceptual and imagery acuity. *Journal of Experimental Psychology: Human Perception and Performance, 7,* 825-832.

Finke, R.A., & Pinker, S. (1983). Directional scanning of remembered visual patterns. *Journal of Experimental Psychology: Learning, Memory, & Cognition, 9,* 398-410.

Friedman, A. (1978). Memorial comparisons without the mind's eye. *Journal of Verbal Learning and Verbal Behavior, 17,* 427-444.

Holyoak, K. (1978). Comparative judgements with numerical reference points. *Cognitive Psychology, 10,* 203-243.

Intons-Peterson, M.J. (1983). Imagery paradigms: How vulnerable are they to experimenters' expectations? *Journal of Experimental Psychology: Human Perception and Performance, 9,* 394-412.

Intons-Peterson, M.J., & White, A.R. (1981). Experimenter naivete and imaginal judgments. *Journal of Experimental Psychology: Human Perception and Performance, 7,* 833-843.

James, W. (1890). *The principles of psychology.* New York: Dover.

Jenkins, J.J. (1981) Can we have a fruitful cognitive psychology? In H.E. Howe (Ed.), *Nebraska symposium on motivation, 1980.* Lincoln, NE: University of Nebraska Press.

Jonides, J., Kahn, R., & Rozin, P. (1975). Imagery instructions improve memory in blind subjects. *Bulletin of the Psychonomic Society, 5,* 424-426.

Kerr, N.H. (1983). The role of vision in "visual imagery" experiments: Evidence from the congenitally blind. *Journal of Experimental Psychology: General, 112,* 265-277.

Koch, S. (1981). The nature and limits of psychological knowledge. *American Psychologist, 36,* 257-269.

Kosslyn, S.M., Ball, T.M., & Reiser, B.J. (1978). Visual images preserve metric spatial information: Evidence from studies of image scanning. *Journal of Experimental Psychology: Human Perception and Performance, 4,* 47-60.

McGuire, W.J. (1973). The yin and yang of progress in social psychology: Seven koan. Journal of Personality and Social Psychology, 26, 309-320.

Marschark, M. (1983). Expectancy, equilibration, and memory. In J.C. Yuille (Ed.), *Imagery, memory and cognition.* Hillsdale, NJ: Lawrence Erlbaum.

Mitchell, D.B., & Richman, C.L. (1980). Confirmed reservations: Mental travel. *Journal of Experimental Psychology: Human Perception and Performance, 6,* 58-66.

Moyer, R. (1973). Comparing objects in memory: Evidence concerning an internal psychophysics. *Perception & Psychophysics, 13,* 180-184.

Neisser, U. (1976). *Cognition and reality*. San Francisco: Freeman.

Neisser, U. (1982). *Memory observed*. San Francisco: Freeman.

Paivio, A. (1971). *Imagery and verbal processes*. New York: Holt, Rinehart, & Winston.

Paivio, A. (1975). Perceptual comparisons through the mind's eye. *Memory & Cognition, 3*, 635-647.

Paivio, A. (1978a). Comparisons of mental clocks. *Journal of Experimental Psychology: Human Perception and Performance, 4*, 61-71.

Paivio, A. (1978b). Mental comparisons involving abstract attributes. *Memory & Cognition, 6*, 199-208.

Paivio, A. (1982, April). *Invited address*. University of North Carolina, Greensboro, NC.

Paivio, A. (1983). The empirical case for dual coding. In J.C. Yuille (Ed.), *Imagery, memory and cognition*. Hillsdale, NJ: Lawrence Erlbaum.

Paivio, A., & Okovita, H.W. (1971). Word imagery modalities and associative learning in blind and sighted subjects. *Journal of Verbal Learning and Verbal Behavior, 10*, 506-510.

Paivio, A., Yuille, J.C., & Madigan, S.A. (1968). Concreteness, imagery, and meaningfulness values for 925 nouns. *Journal of Experimental Psychology Monograph, 76*(1, Pt.2).

Petrinovich, L. (1979). Probabilistic functionism: A conception of research method. *American Psychologist*, 373-390.

Pylyshyn, Z.W. (1979). The rate of "mental rotation" of images: A test of a holistic analogue hypothesis. *Memory & Cognition, 7*, 19-28.

Pylyshyn, Z.W. (1980). Computation and cognition: Issues in the foundations of cognitive science. *The Behavioral and Brain Sciences, 3*, 111-169.

Richman, C.L., Mitchell, D.B., & Reznick, J.S. (1979). Mental travel: Some reservations. *Journal of Experimental Psychology: Human Perception and Performance, 5*, 13-18.

Shepard, R.N. (1978). The mental image. *American Psychologist, 33*, 125-137.

Shepard, R.N., & Chipman, S. (1970). Second-order isomorphism of internal representations: Shapes of states. *Cognitive Psychology, 1*, 1-17.

Shepard, R.N., & Metzler, J. (1971). Mental rotation of three-dimensional objects. *Science, 171*, 701-703.

Steiger, J.R., & Yuille, J.C. (1983). Long term memory and mental rotation. *Canadian Journal of Psychology, 37*, 367-389.

Stent, G.S. (1975) Limits to the scientific understanding of man. *Science, 187*, 1052-1057.

Wundt, W. (1902). *Outline of psychology* (C.H. Judd, Trans.). New York: G.E. Stechert.

Yuille, J.C. (1983). *The ultimate failure of cognitive psychology*. Unpublished manuscript. University of British Columbia.

Yuille, J.C., & Marschark, M. (1983a). Imagery effects in memory: Theoretical interpretations. In A. A. Sheikh (Ed.), *Imagery: Current theory, research, and application*. New York: Wiley and Sons.

Yuille, J.C., & Marschark, M. (1983b). *Computational psychology: A reincarnation of behaviorism?* Unpublished manuscript. University of British Columbia.

Yuille, J.C., & Steiger, J.H. (1982). Nonholistic processing in mental rotation: Some evidence. *Perception & Psychophysics, 31*, 201-209.

Zimler, J., & Keenan, J.M. (1983). Imagery in the congenitally blind: How visual are visual images? *Journal of Experimental Psychology: Learning, Memory, and Cognition, 9*, 269-282.

CHAPTER **9**

The Neuropsychology of Imagery

David F. Marks

I. INTRODUCTION

On what basis can one hope to discover how the complex mental function of imagery is represented in the neural tissue? We know that the primary visual area is 17, the primary auditory area is 41, and the primary somatosensory areas are 1, 2 and 3. Secondary areas which integrate sensory information are 18 and 19 for vision, 22 and 42 for audition, and 5 and 7 for somaesthesis. These three sensory areas are further integrated in areas 39 and 40, which are designated the "parietal association cortex." This tertiary area determines the dominancy for speech and is most important for cognitive functions because it integrates sensory information from all sources.

The speech areas have received such study and have been localized in the majority of cases in the left or "dominant" hemisphere. The

225

so-called "minor" right hemisphere has received less study but seems to have a role in spatial ability, the body schema, face perception, musical perception and other nonverbal mental functions. Many researchers have assumed that the right hemisphere is the "imagery hemisphere," neatly localizing speech and language "in" the left hemisphere and imagery "in" the right. The illogicality of the conventional argument for right-hemispheric localization of imagery can be seen in a recent review by Ley (1983). Ley suggests the following syllogism in favor of right-hemispheric localization:

The right hemisphere subserves spatial perception. Imagery is integral to spatial-perceptual abilities. Thus, the right-hemisphere subserves imagery processes.

As Ley acknowledges, such overly simple reasoning is bound to be specious and the bold attempt Ley makes to provide the necessary empirical "bootstrapping" is somewhat unconvincing. In this chapter the evidence for a quite different theory of imagery representation will be presented.

In the first part of this chapter, some evolutionary precursors of imagery will be described and a case for imagery as a system of representation more ancient than language will be made. Possible neurological mechanisms for waking and hypnagogic imagery will then be described together with some recent speculations concerning a related, but opposite, mechanism for the dream imagery of sleep. Clinical neuropsychological studies suggesting that imagery is topologically represented across both hemispheres of the cortex will then be presented.

II. IMAGERY IN EVOLUTION

1. Imagery and Consciousness

"Reality" is a construction manufactured by the nervous system out of a long and complex sequence of sensory input. Although one's construction of reality is socially and culturally determined, the foundation for one's model of reality is the nervous system. Perception could be said to provide the bricks and imagery the mortar. Language provides the means for communicating one's idea of reality to another. Thus, while human beings reared in isolation will not learn language, an innate capacity for imagery will create a sense of reality unimpeded by the absence of language.

In agreement with Jerison's (1973) analysis of the *Evolution of the Brain and Intelligence*, imagery clearly must have preceded language as the means of representing and modeling the real world. To quote Jerison (1973):

The quality of language that makes it special is less its role in social communication than its role in evoking cognitive imagery ... it was this kind of capacity that was evolving in the early hominids. (We need language to tell a story much more than to give directions for an action.)

In this view, language is an instrument for storytelling, for conveying to others an imagined or constructed scenario, past, present or future. In Jerison's (1973) words:

Individuals capable of constructing elaborate multisensory "real" words might construct a reality that seems more fundamental than the immediate information from the senses. The capacity for imagery, in which one manipulates a possible real world in one's imagination, must early have led the hominids, by the time these capacities were well developed, to reach an appreciation of a past prior to one's lifetime and a sense of future after one's death (p. 429).

Jerison's evolutionary theory postulates that early language contributed to the capacity for imagery among early hominids. For 99% of their evolution the hominids lived in nomadic groups as hunter-gatherers covering enormous ranges of land. Unlike other predators such as wolves, hominids had a weak olfactory sense, and the marking capacity through urine and scent glands was lacking. In Jerison's theory, language provided an equivalent system of extending the range of the hunt by marking it vocally via naming relevant features of the terrain. Vocalization and hearing, therefore, made up for the lack of olfactory sense in the production of an image of the environment. Through vocalization, one member of a nomadic group could evoke images in the mind of another. Thus, in Jerison's theory, speech and language evolved as a mechanism for sharing imagery.

But is it meaningful to look for evidence of imagery prior to the evolution of hominids? Imagery has generally been conceptualized as an aspect of human consciousness. Unfortunately, the term "consciousness" has all kinds of connotations and meanings. Armstrong (1979) describes three of those meanings: minimal, perceptual, and introspective. Armstrong starts with a description of unconsciousness. This is a total lack of mental activity of any kind: no perception, no sensations, no feelings, no desires, no thoughts, no images. *Minimal consciousness* occurs if there is some mental activity, for example, a dream or a solution to a problem occurring upon awakening but not present before sleep. *Perceptual consciousness* is literally the type of mental activity which occurs during perception - an awareness of the environment and bodily states. *Introspective consciousness* occurs when one is aware of the mental. This is almost certainly a late evolutionary development. According to Jaynes (1976) consciousness in this sense occurred with a hypothesized "breakdown of the bicameral mind" about 3000 years ago, and, as both Armstrong and Jaynes have pointed out (like many before them), introspective consciousness is bound up with a very special characteristic -

the concept of self: "Without introspective consciousness we would not be aware that we existed - the self would not be self to itself" (Armstrong, 1979, p. 241).

If imagery is defined as a part of the introspective consciousness, then it would not be possible in animals. However, if we allow that imagery can be part of perceptual consciousness (e.g., in the form of expectancies) and of minimal consciousness such as that which occurs in dreaming, hallucinations, epileptic seizures and similar happenings, then a case for imagery in animals can be made.

2. Eidetic Imagery

Perhaps the most primordial kind of imagery is the apparent eidetic imagery of insects (Collett & Cartwirght, 1983). Insects show a very accurate ability to navigate their environment. Some wasps, for example, dig burrows in sand to lay eggs and fly miles away, returning much later to feed the larvae. It appears that such insects learn the size of landmarks near the nesting site by taking "retinal snapshots" of them. Desert ants appear to use a similar process, as illustrated by the experiments of Wehner and Räber (1979). A nest opening of an ant colony was marked by two small black cylinders placed on either side of the nest. After allowing the ants to become accustomed to the presence of the landmarks, both landmarks and ants were taken to a new area and the ants' search patch were recorded. When the constellation of landmarks was similar to that used in training, ants searched midway between them, where the nest ought to have been. They did likewise when the distance and size were both doubled so that the landmarks viewed from the midway point had the accustomed appearance. However, if size or distance were altered disproportionately, the ants searched very close to one or the other landmark. Collett and Cartwright (1983) propose a photographic analogy for the insect's search process which they refer to as a "snapshot." The hypothesized snapshot of an insect is only matched to the retinal image formed by the landmarks when the insect arrives at the original nest or food source.

Studies of navigation in birds and fish also suggest the existence of powerful innate homing mechanisms. However, the need to postulate imagery for such behavior is less clear than it is for insect navigation. The sun, the stars, and geomagnetism all provide compass information for bird migration and landmarks do not seem to be used as they are for insects. Salmon ranches rear hundreds of thousands of young salmon and then free them into the ocean in the hope that a certain percentage will later come "home," But, again, it would be too big a jump to suggest that imagery is involved. O'Keefe and Nadel (1978) have persuasively argued that cognitive maps for spatial representation are an automatic function of the hippocampus. No mental imagery is required for the acquisition of such maps, merely the curiosity to explore the environment, and no mental imagery is necessary for their application.

3. Dreams

In turning to the study of sleep and dreams, there is evidence of REM sleep in all mammalian species, and some birds. Careful studies confirming the occurrence of REM sleep have been conducted with the elephant, chimpanzee, baboon, whale, pig, sheep, monkey, rat, mouse, cat, bat, dog, donkey, guinea pig, the opposum, kangaroo and many others (Meddis, 1977). The issue of whether dreams occur in association with REM sleep in animals is a moot point. Perhaps, as Meddis (1977) suggests, the dream characteristics of REM sleep have been over-emphasized, while its neurological foundation has been practically ignored. Meddis's theory of REM or "active" sleep, as he terms it, is that it evolved and descended from reptilian sleep. For warm-blooded animals, active sleep (AS) is inappropriate, as the reflexes which maintain normal body temperature during waking are inoperative (Permeggiana & Rabini, 1970). Hence, Meddis suggests that active sleep had to be restricted to brief, well-spaced episodes and non-REM or quiet sleep evolved to fill in the gaps.

The notion that animals are capable of dreaming has long been espoused. Lucretius in the second century A.D. observed muscular activity in sleeping horses and assumed the movements were related to the horses' dreams (Hartmann, 1967). Common observation of our favorite pet cats and dogs show facial expressions and paw movements congruent with possible dream activity such as running or hunting. Evarts (1962) observed that neurones in the cat's occipital cortex were highly active during REM sleep. Vaughn (1964) conditioned monkeys to press a bar whenever they saw a visual image on a screen in front of them. The response was not conditioned specifically to particular visual images but to a large variety of them. During both sensory deprivation and REM sleep, but not in non-REM sleep, the monkeys would suddenly begin to press the bar. Hence, in the cat and the monkey visual imagery does seem to accompany REM sleep and the dream hypothesis receives support.

4. Hallucinations

A further source of evidence of imagery in animals is their behavior following ingestion of hallucinogens. Siegal and Jarvik (1975) have recounted how boars, porcupines and gorillas in the jungles of Gabon and the Northern Congo dig up and eat the roots of *Tabernanthe iboga*, an hallucinogenic plant containing ibogaine. Some Indian groups in Central Mexico apparently abstain from *Cannabis* because local monkeys continually raid the feed on the young plants. This has led to a long-held belief there that cannabis is "food fit only for animals." Laboratory studies of many species have trained animals to report perceptual events and then have substituted an hallucinogen treatment. Siegal (1969) has conducted a number of studies of this kind with pigeons and gained positive results. Siegal and Jarvik's (1975) comprehensive review of the literature relevant to the question, "Do animals hallucinate?" led them to conclude: "Yes, they do. We all do" (p.104).

III. NEUROLOGICAL MECHANISMS

1. Waking Imagery

Neuropsychological research with adult humans supports the view that imagery provides a fundamental mechanism for representing the external world, or what we fondly term "reality." The neuropsychological theory of Hebb (1949, 1980) provides a useful framework for reviewing much of the relevant research. Hebb suggested that imagery and thought are the activity of a neural holding mechanism which he termed the "cell assembly." Hebb developed this idea after seeing Lorente de No's (1938) description of neural networks which contained closed loops or re-entrant paths.

Figure 1 gives an example of how these reverbatory circuits might operate. A single fiber from outside the system excites four neurons, A, B,

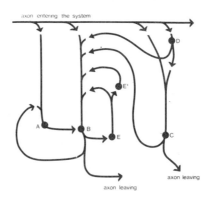

Figure 1. Hebb's theory of reverberatory circuits in the cortex. A fiber from outside the system excited four neurons, A. B, C. and D. A excites B. which re-excites A as well as exciting E. E re-excites B directly and also indirectly via E'. Two neurons send axons to produce effects outside of the system. (Adapted from a drawing in Hebb, 1980).

C, and D. A excites B, which re-excites A, as well as exciting E. E re-excites B directly and also indirectly via E', and so on. Two of the four neurons send their axons out of the system so that the internal activity will have effects elsewhere. When an input enters the system, instead of being transmitted immediately on to a motor path, or extinguishing itself, the signal can be held long after the original sensory stimulation has ceased.

Such cortical holding activity can be set off centrally and so a cell-assembly originally activated by percpetual activity can later be re-activated through its connection with other cortical assemblies. No sensory inputs are necessary for the central "perceptual" cell-assemblies to set off a train of thought.

A major postulate in Hebb's theory was the assumption that the interconnectivity of neurones can be strengthened. This strengthening was assumed by Hebb to occur whenever a synapse was active at the time the postsynaptic neuron discharges. Synaptic strengthening provides a neural basis for association. If several inputs to a cell occur simultaneously, they will be sufficient to activate the cell and all the active synapses will be strengthened. Later on, each of those input cells will find it easier to fire the target cell and may do so without the existence of the other associated inputs. When Hebb's theory was originally proposed in 1949, it was untested and untestable. Eventually, in the 1970's, Bliss and Lomo (1973) were able to show that long-lasting changes could be produced in the strength of synapses in the brain of adult mammals. The technique they used delivered a pulse of electricity to a neural pathway and the magnitude of the response in the area where the pathway projects was recorded. Once a baseline had been established for the evoked response due to the test pulse, a high frequency train of pulses was delivered to the neural pathway. The test pulses were then repeated at some later time and the evoked response was again recorded. Bliss and Lomo discovered that long-lasting increases occurred in the hippocampus following pulse trains in the perforant path. Bliss and Lomo termed this phenomenon "long term potentiation" (LTP). It provided a possible synaptic strengthening mechanism postulated by Hebb and, potentially, a neural basis for learning. The major obstacle for the establishment of LTP as a neural mechanism for learning is the provision of evidence of LTP accompanying behavioral evidence of learning.

As pointed out by Bugelski (1982, see Chapter 4), a fundamental similarity could exist in the neural mechanisms underlying imagery and learning. Imagery represents the activity of neural processes when excited by associated sensory or central processes; learning represents the associating together of sensory inputs with one another, with entirely central processes, or with motor processes. Hence, LTP could provide the neural mechanism for both learning and imagery. Ahsen (1978) postulated this same connection based on his findings from eidetic imagery.

2. Hypnagogic Imagery

In a classic study of the decomposition and fusion of mental images, Naruse and Obonai (1953) reported the following example. Obonai, while lying half-awake in his bed, reported seeing an image of a frog on something like a tree. The color of the frog's back was bluish green, and the form was a little bigger and more grotesque than a tree frog. Meanwhile, an image of a *phyllostachyomitis* growing in the thick growth of plants in the garden appeared shiftily. Soon after, it changed to an image of a radish planted in the garden at the back of the house. The imagined radish revealed a white skin about one-third of the total root-length above the earth, and he also saw leaves. At first, he didn't notice the tight clenching of his left arm with his right hand until he realized that it bore some resemblance to the experience of touching the radishes which he had pulled out with his right hand in the garden a few days previously. Releasing his hand, the radish root disappeared, and the radish leaves only remained. Again clenching his arm, at first something like yellowing leaves of the radish appeared, and in the meantime many radish roots appeared. Then the radish roots and leaves all disappeared.

Naruse and Obonai (1952) gave this experience the following interpretation: "It is raining heavily all evening long, and still at bed time a pouring rain showers the roof tiles. He can hear the noisy rain showering the bamboo blades. When he hears the noise, he sees also the images of the trees or bamboos which are associated with the noise. The image frog is a tree frog which he saw a few days ago on an oak leaf. On the noon of that day, he observed a toad creeping about in the garden for some time. Then the grotesque form of the imagined frog is perhaps due to a fusion of the memory image of the frog and that of the toad. The images of the frog and bamboos appear for a long time, because of the heavy and endless rain. The radish images are responsible chiefly for the touch sense of the hand and arm" (p. 25).

Naruse and Obonai (1953) concluded their analysis of the hypnagogic state with the statement that *as a result of conditioning*: "When one sensory stimulus is given to a subject in a drowsy state, images of other objects associated with it often appear" (see also Chapter Three). Thus, in agreement with Hebb, Bugelski, and Naruse, we can conclude that imagery is evoked as a result of learned associations between cell-assemblies. Whether the eliciting stimulus is external (and possibly subliminal) or central (and possibly not conscious) imagery results from an associated neural excitation which is a product of learning. The process of LTP is possibly the underlying mechanism.

3. Dreams

Is it possible that one simple mechanism - association via synaptic strengthening - could explain all of imagery's multifarious characteristics? While ordinary memory, imagination, hypnagogic imagery and even eidetic imagery might all conceivably result from the triggering of associations between experiences and ideas, what about dreams? Surely dream imagery cannot also result from associations formed in everyday waking reality. Freud tried to explain dream imagery as symbolic expression of unconscious drives. Apart from the obvious subjectivity of dream interpretation, there are many other problems with the Freudian theory of dreams. Three of the major ones are : (1) although we dream for 1-2 hours every night, we rarely recall any dreams; (2) REM sleep has been found in all mammals and in many birds. If we believe that all the REM spleep is dream sleep, it stretches one's credulity to assume that all of these species have such well-developed super-egos that they have to dream; (3) huge amounts of dream sleep have been found in young animals and babies and also in the in utero fetus.

Crick and Mitchison (1983) have proposed a new theory of dream sleep which does away with much of the old mystery:

The cortical system. . .can be regarded as a network of interconnected cells which can support a great variety of modes of mutual excitation. Such a system is likely to be subject to unwanted or 'parasitic' modes of behavior, which arise as it is disturbed either by the growth of the brain or by modifications produced by experience. We propose that such modes are detected and suppressed by a special mechanism which operates during REM sleep and has the character of an active process which is, loosely speaking, the opposite of learning (p. 111).

Crick and Mitchison call this process 'reverse learning' or 'unlearning'.

During cortical growth and alteration as result of experience, Crick and Mitchison have predicted that a number of undesirable or "parasitic" modes of activity would be expected to occur in complex neural networks which are being "wired" together. As associations are acquired in various cell-assemblies, the information is not necessarily stored in a highly specific location for each particular item. Information is disturbed over many synapses and is therefore robust in that it won't be lost if a small number of synapses are deleted or added. Information is also superimposed as a given synapse may be involved in several areas or topics of information. Any one neurone may be connected with many thousands of others.

A well-formed cell-assembly can be trained so that a given input will produce an appropriate output. Certain properties of a well-formed cell-assembly are described by Crick and Mitchison (1983) as follows: (1) *Completion*: The memory is content addressable. Given part of the input, it can produce the whole of the output; (2) *Classification*: Given an input which is related to several of a cell-assembly's associations, an output can be produced which combines the common features of its input.

One problem with such neural networks, Crick and Mitchison argue, is their ability to become overloaded when an attempt is made to store too many different patterns or if there is too much overlap. Such overloading leads to poor performance which can take any of the following forms: (1) the network may produce too many far-fetched or bizarre associations or fantasies; (2) the network may tend to produce the same perseverative state regardless of the specific input - it becomes "obsessed"; (3) networks which contain feedback loops may respond to inappropriate signal levels which normally should elicit no response. The system as a whole "hallucinates."

How has evolution coped with these apparently inevitable consequences of growing, complex neural networks? How are bizarre fantasy, obsession, and hallucination successfully avoided in the normal waking state? Crick and Mitchison ingeniously suggest that the system would require the following actions: (1) major inputs and outputs to be switched off so the system is largely cut off from outside influence; (2) random activation from inside should then be switched on to dampen down or weaken the synapses which are becoming falsely strengthened purely because of this overloading.

REM sleep would appear to have the required characteristics. The cortex is periodically but widely stimulated by the brain stem by what are known as PGO (potine-geniculate bodies - occipital cortex) spikes. Hobson and McCarley (1977), following Jouvet (1962), suggest that the PGO waves cause the REMs themselves and also dreams.

In Crick and Mitchison's theory, PGO spikes are assumed to weaken the synapses that tend to be involved in the temporary effects of growth and new experience, the causes of the undesired modes of activity. Hence, dreams would have the opposite effect to learning: We would dream in order to forget.

Hopfield, Feinstein and Palmer (1983) report evidence from neural modeling that false memories produced in growing neural networks can be successfully suppressed by an unlearning mechanism such as that described by Crick and Mitchison. Crick and Mitchison's unlearning theory of dreams, while highly speculative, is compatible with a large amount of empirical data: (1) it explains the observation of the large amount of

REM-sleep found in young, developing nervous systems; (2) it explains the hallucinatory nature of dreams; (3) it explains why we seldom remember dreams; and (4) it explains the relationship which exists between brain size and REM time: the larger and neuronally more complex the neocortex, the greater the amount of REM sleep.

Crick and Mitchison suggest that without REMs, and the dreams that are generated by PGO spikes, evolution could not have produced such a highly successful neocortex. A breakdown of the reverse learning mechanism would produce the phenomena of hallucinations, delusions, and obsessions, some of the major symptoms of schizophrenia.

IV. NEUROPSYCHOLOGICAL EVIDENCE

To sum up our thesis thus far, it has been argued that imagery is an older system of modeling the world than language. We have outlined some ideas from Hebb concerning the neural mechanisms for imagery, and the supporting evidence, and the ideas from Crick and Mitchison concerning dream mechanisms and the relevant supporting evidence.

What evidence is there in clinical neuropsychology to support the position that imagery is independent of language as a mode of conscious representation? I will briefly outline the evidence in the following areas: the study of lesion effects, cortical stimulation, commissurotomy and EEG research.

1. Unilateral Neglect

The first group of studies on lesions causing unilateral neglect have been conducted at the University of Milan (Bisiach, Capitani, Luzzati, and Perani, 1981) and at the University of Auckland (Ogden, 1983). Lesions in the parietal lobes, frontal and cingulate cortex, and basal ganglia have all been associated with hemi-neglect. This is indicated by an unawareness of stimuli in the side of space contralateral to the lesion. While Ogden (1983) has shown that hemi-neglect occurs for both right and left hemisphere damage, the Milan group found that the neglect effect occurred not only in perception but also in imagery. The task they used was to ask their subjects to visualize a highly familiar target site, the cathedral square in Milan. This place has the advantage of not only being reasonably symmetrical, but of having approximately equal numbers of features on each side.

The subjcets were asked to imagine themselves facing the front of the cathedral from the opposite side of the square. Once the description was completed, the subject was asked to imagine the square again, this time imagining the vantage point to be the central entrance of the cathedral. The procedure was then repeated and the subject was specifically prompted to describe the left and right sides of the square from the two opposite vantage points.

Subjects with lesions in the right hemisphere displayed an inability to describe image content for the left half of the imaged field. When the vantage point was reversed, things which the subjects had described on the previous good side were now omitted and things which had previously been omitted (the bad side) were now described. The same pattern of results has since been replicated by Ogden (1985).

These data suggest that the pattern of brain activity underlying imagery is to some extent isomorphic with the object being represented. Imaginal space is apparently topologically structured across the two hemispheres. Bisiach's group believe that the locus for this representation is the association cortex of the parieto-tempero-occipital junction. This could well be the case in the nondominant hemisphere.

Ogden's (1983) data suggests that the representation of visual space is relatively anterior in the left hemisphere and posterior in the right. Perhaps this is not surprising, considering Wernicke's area in the left cerebral hemisphere occurs at the locus designated for imagery by Bisiach.

2. Loss of Imagery

Another study from the Milan group (Basso, Bisiach, & Luzzatti, 1980) reported on an individual who totally lost his imagery ability following a lesion in the left occipital region after a stroke. The patient was given the "Cathedral Square" test, and was found incapable of naming a single item in the square part from the cathedral itself, and the monument in the center of the square. He could not form images of faces, including his wife's, although he *knew* she was "small, grey-haired with almond-shaped eyes." He stated that he used to enjoy "very real imagery," including hypnagogic images of game before or after a day's hunting, his major pastime. "Now," he stated, "when I am lying sleepless in my bed I cannot even think!" The loss of imagery was not confined to the visual modality-the patient was unable to imagine odors, tastes, sounds and he was no longer able to hum tunes.

Soon after the stroke, there was dysphasia and the patient's reading and writing were impaired, but four months later these problems had disappeared. Naming pictures of familiar objects and place-naming were impaired for much longer and loss of visual imagery seemed permanent. The now imageless patient could find his way from his home to the hospital and back, but could not describe the journey. He was also unable to state the route taken by the tram he had driven for many years. Several other similar cases have been reported in the literature (e.g., Brain, 1954).

On the basis of the other research already described, it would be incorrect to jump to the conclusion that imagery for this patient was "in" his left occipital cortex. A much more probable explanation of the disorder is that a disconnection occurred between brain processes mediating language and those mediating imagery. Hence, the patient's "introspective consciousness" had become oblivious to mental imagery. In Geschwind's (1965) words: "If a part of the brain is fully disconnected from the speech area, it will not be possible for the speech area to give an account of what goes on in that part of the brain" (p. 638).

One wonders whether the disconnection syndrome described originally by Geschwind (1965) might naturally occur in varying degrees subclinically, giving rise to a huge variation in the reported imagery which exists in the population (Marks, 1983). As Geschwind has pointed out, every brain is diffferent and variations in connectivity could well have implications for psychological functioning, particularly in the area of conscious imagery experience.

3. Cortical Stimulation

The coritcal organization of imagery has also been revealed in studies of brain stimulation (e.g., Penfield & Perot, 1963). Flashbacks of experientiàl hallucinations occurring in epileptic seizures and other memories may be evoked artificially by electrical stimulation of the cortex. The areas found by Penfield and Perot (1963) to yield experiential responses fell mostly within the temporal lobes and, although there were more on the right side, they occurred in both hemispheres. The posterior speech area (Wernicke's) did not yeild any experiential responses. Auditory responses were limited to the lateral and superior surface of the first temporal convolutions. Voices were most common (46 cases out of 66) and music sound (17 cases out of 66).

Visual experiential responses occurred in 38 cases. In 23 cases, a person or persons doing something was seen; in 10 cases, scenes: and in 5 cases, objects. In 22 cases, combined auditory and visual responses were obtained, many of them musical, such as somebody singing or playing a piano.

Looking at the overall summary (see Figure 2) one is struck by the fact that the speech area seems to have displaced the topical organization of imagery in the so-called dominant hemisphere. On the non-dominant side, where there is no speech cortex, the homologous area is devoted to imagery and is much larger than the visual imagery area available on the dominant side. One suspects, on the basis of Ogden's (1983) hemi-neglect data, that visual imagery in the dominant hemisphere may also be mediated anteriorly in positions unavailable for stimulation in the Penfield

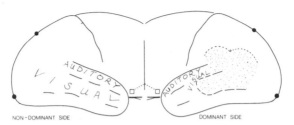

Figure 2. A summary of Penfield's data on the distribution of experiential responses of imagery across the cortex. (Adapted from a drawing in Penfield & Perot, 1963)

series. However, even in Penfield's data, one can see that imagery responses are produced by both sides of the cortex, while speech is produced by only one.

4. Commissurotomy

Studies of imagery in patients with a split brain are few. However, data produced by Greenwood, Wilson and Gazzaniga (1977) show that such patients do report dreams. Hoppe (1977) even found dreams occurring in a patient with a right hemispherectomy. These data establish that the left hemisphere does have the capacity for producing imagery.

5. EEG Measures

Another avenue for the neuropsychological investigation of imagery has been studies using the EEG. For 40 years it has been assumed that imagery causes EEG alpha activity to attenuate (Golla, Hutton, & Walter, 1943). Recent studies have focused on the idea that imagery might be a right hemisphere process. The evidence is, to say the least, equivocal. Most of the research showing asymmetry in alpha suppression has confounded imagery with a reduction of verbal thinking and so the results are impossible to interpret (Ehrlichman & Barrett, 1983). As pointed out by Ehrlichman and Barrett (1983), an experimental design is needed in which these two processes (verbal and imagery) do not covary. Only two studies appear to have achieved this, and both failed to find any evidence of asymmetry (Barrett & Ehrlichman, 1982; Haynes & Moore, 1981).

Barrett and Ehrlichman (1982) asked questions designed to induce either high or low imagery activity. Both question types required some verbal responses (e.g., high imagery: "What does your TV set look like"?; low imagery: "What does this mean? Beggars can't be choosers."). Bilateral EEGs were recorded and also eye movements. Supporting previous research that eye movements are reduced during waking imagery (Antrobus, Antrobus & Singer, 1964; Marks, 1973; Weiner & Ehrlichman, 1976), high-imagery questions produced significantly less eye movement.

Also, bilateral alpha activity was significantly lower in high-imagery than in low-imagery items. However, the EEG asymmetry index showed no difference in the balance of activation between the high - and low-imagery conditions. Although the results support the idea of alpha blocking as a correlate of imagery functioning, there was no asymmetry in the blocking response.

A new EEG technique for the study of mental functions is EEG topographical mapping. Mathematical interpolation by high-speed computer is used to generate a mapping of electrical activity across the cortex based upon the output of an array of electrodes. EEG topographical analysis reduces a large quantity of data into an easily perceivable form. EEG topographic mapping has yet to be performed for imagery tasks, although a study is currently in progress in collaboration with Professors Kenichi Uemura of Hamamatsu University School of Medicine and Jiro Tatsuno of the National Defence Medical College in Japan (Marks, Uemura, & Tatsuno, 1984).

V. CONCLUSION

In this chapter it has been argued that imagery is an older system of modeling the world than language. There is evidence of eidetic-like imagery mechanisms in insects, and of dreaming and drug-induced waking hallucinations in birds and mammals. Neurological mechanisms for imagery in the waking, hypnagogic and dream states have been proposed. These consist of neural networks with synapses of modifiable strength. These networks may be excited by sensory input, central activity, and by PGO spikes during REM sleep. Neuropsychological data from a number of sources suggest that imagery is topologically represented across both hemispheres of the cortex. Antero-posterior asymmetry of imagery representation exists between the two hemispheres associated with the lateral asymmetry of the speech and language centers.

REFERENCES

Ahsen, A. (1978). Eidetics: Neural-experiential growth potential for the treatment of accident trauma, debilitating stress conditions, and chronic emotional blocking. *Journal of Mental Imagery, 2,* 1-22.

Antrobus, J.S., Antrobus, J.S., & Singer, J.L. (1964). Eye movements accompanying daydreaming, visual imagery, and thought suppression. *Journal of Abnormal & Social Psychology, 69,* 244-252.

Armstrong, J. (1979). Consciousness. In *Brain and Mind* (CIBA Foundation). Amsterdam: Excerpta Medica.

Barrett, J., & Ehrlichmann, H. (1982). Bilateral hemispheric alpha activity during visual imagery. *Neuropsychologia, 20,* 703-708.

Basso, A., Bisiach, E., Luzzatti, C. (1980). Loss of mental imagery: A case study. *Neuropsychologia, 18,* 435-442.

Bisiach, E., Capitani, E., Luzzatti, C., & Perani, D. (1981). Brain and conscious representation of outside reality. *Neuropsychologia, 19,* 543-552.

Bliss, T.V.P., & Lomo, T. (1973). Long-lasting potentiation of synaptic transmission in the detente area of the anaesthetized rabbit following stimulation of the perforant path. *Journal of Physiology, 232,* 331-356.

Brain, R.W. (1954). Loss of visualization. *Proceedings of the Royal Society of Medicine, 47,* 288-290.

Bugelski, B.R. (1982). Learning and imagery. *Journal of Mental Imagery, 6,* 1-92.

Collett, T.S., & Cartwright, B.A. (1983). Eidetic images in insects. Their role in navigation. *Trends in Neuroscience, 6,* 101-105.

Crick, F., & Mitchison, G. (1983). The function of dream sleep. *Nature, 304,* 111-114.

Ehrlichman, H., & Barrett, J. (1983). Right hemisphere specialization for mental imagery: A review of the evidence. *Brain & Cognition, 2,* 55-76.

Evarts, E. (1962). Activity of neurons in visual cortex of cat during sleep with low voltage fast EEG activity. *Journal of Neurophysiology, 25,* 812-816.

Geschwind, N. (1965). Disconnexion syndromes in animals and man. *Brain, 88,* 237-294, 585-644.

Golla, F., Hutton, E.L., & Walter, W.G. (1943). The objective study of mental imagery. I. Physiological concomitants. *Journal of Mental Science, 89,* 216-223.

Greenwood, P., Wilson, D.H., & Gazzaniga, M.S. (1977). Dream report following commissurotomy. *Cortex, 13,* 311-316.

Hartmann, E. (1967). *The biology of dreaming.* Springfield, IL: Charles C. Thomas.

Haynes, W.O., & Moore, W.H. (1981). Sentence imagery and recall: An electroencephalographic evaluation of hemispheric processing in males and females. *Cortex, 17,* 49-62.

Hebb, D.O. (1949). *Organization of behavior.* New York: John Wiley.

Hebb, D.O. (1980). *Essay on mind.* Hillsdale, NJ: Lawrence Erlbaum Associates.

Hobson, J.A., & McCarley, R.W. (1977). The brain as a dream state generator: An activator synthesis hypothesis of the dream process. *American Journal of Psychiatry, 134,* 1335-1348.

Hopfield, J.J., Feinstein, D.I., & Palmer, R.G. (1983). "Unlearning" has a stabilizing effect in collective memories. *Nature, 304,* 158-159.

Hoppe, K. (1977). Split brains and psychoanalysis. *Psychoanalytic Quarterly, 46,* 220-244.

Isaac, A. (1985). Imagery differences and mental practice. In D.F. Marks & D.G. Russell (Eds.) *Imagery 1.* Dunedin: Human Performance Associates.

Jaynes, J. (1976). *The origin of consciousness in the breakdown of the bicameral mind.* Boston: Houghton Mifflin Company.

Jouvet, M. (1962). Recherches sur les structures nerveuses et les mecanisms responsables des differentes phases du sommeil physiologique. *Archives of Italian Biology, 100,* 125-206.

Ley, R.G. (1983). Cerebral laterality and imagery. In A.A. Sheikh (Ed.), *Imagery: Current theory, research, and application.* New York: John Wiley.

Lorente de No, R. (1938). Analysis of the activity of the chains of internumial neurons. *Journal of Neurophysiology, 1,* 207-244.

Marks, D.F. (1973). Visual imagery differences and eye movements in the recall of pictures. *Perception & Psychophysics, 14,* 407-412.

Marks, D.F. (1983). Mental imagery and consciousness: A theoretical review. In A.A. Sheikh (Ed.) *Imagery: Current Theory, research and application.* New York: John Wiley.

Marks, D.F., Uemura, K., & Tatsuno, J. (1984, September). *EEG topographic analysis of imagery.* Paper presented at the International Congress of Psychology, Acapulco, Mexico.

Meddis, R. (1977). *The sleep instinct.* St. Lucia, Queensland: University of Queensland Press.

Naruse, G., & Obonai, T. (1953). Decomposition and fusion of mental images in the drowsy and post-hypnotic hallucinatory state. *Journal of Clinical & Experimental Hypnosis, 1*, 23-41.

Ogden, J.A. (1983, August). *Antero-posterior interhemispheric differences in the loci of lesions producing visual hemineglect.* Paper presented at the meeting of the Australasian Winter Conference on Brain Research, Queenstown, New Zealand.

Ogden, J. (1985). A disorder of visual imagery as a consequence of unilateral brain damage. In D.F. Marks & D.G. Russell (Eds.). *Imagery 1.* Dunedin: Human Performance Associates.

O'Keffe, J., & Nadel, L.(1978). *The hippocampus as a cognitive map.* New York: Oxford University Press.

Penfield, W., & Perot, P. (1963). The brain's record of auditory and visual experience. *Brain, 86*, 595-696.

Siegal, R.K. (1969). Effects of *Cannabnis sativa* and lysergic acid diethylamide on a visual discrimination task in pigeons. *Psychopharmacologia, 15*, 1-8.

Siegal, R.K., & Jarvik, M.E. (1975). Drug-induced hallucinations in animals and man. In R.K. Siegel & L.J. West (Eds.), *Hallucinations: Behavior, experience, and theory.* New New York: John Wiley.

Vaughn, C. (1964). *The development and use of an operant technque to provide evidence for visual imagery in the rhesus monkey under sensory deprivation.* Unpublished doctoral dissertation, University of Pittsburgh.

Wehner, R., & Räber, F. (1979). Visual spatial memory in desert ants, Cataglyphis bicolor (Hymenoptera: Formicidae). *Experientia, 35*, 1569-1571.

Weiner, S.L., & Ehrlichman, H. (1976). Ocular motility and cognitive process. *Cognition, 4*, 31-43.

CHAPTER 10

Toward a New Structural
Theory of Image Formation

David F. Marks

Historically, five major "schools" or "paradigms" have been concerned in one way or another with the psychological investigation of imagery. Each has promoted its own particular methodological solutions to the problem of acquiring knowledge in psychology, and the analysis of imagery has played a role in each of these formulations. These are the experimental approaches (Fechner, Wundt, and Galton), recently revived as "cognitive" psychology, psychoanalytic-dissociationist (Freud, Charcot, and Janet) behaviorist (Watson, Holt, and Skinner) neuropsychological (Flourens, Luria, and Hebb), and Piagetian development (Piaget and Inhelder).

While each paradigm has made a contribution to our understanding of imagery, large gaps remain in our knowledge resulting from a general

This chapter is an adaptation of two articles published in the *Journal of Mental Imagery* in 1984 and 1985. These were entitled "The New Structural Approach to Image Formation, Psychophysiology, and Psychopathology" and "Imagery Paradigms and Methodology" respectively. I would like to acknowledge many helpful discussions with Akhter Ahsen and Peter McKellar.

neglect of *image structure* and of its relationship to the individual's past and future psychological development. Methods in science are a function of theoretical aims, and in this chapter I wish to clarify this relationship with special reference to imagery. Following a brief review of traditional approaches, a new paradigm which combines the most helpful elements of all of the others will be described. This approach is originally based upon clinical researches, in particular those conducted by Ahsen (1965, 1972, 1977, 1982).

The *New Structuralist* paradigm provides the theoretical rationale and methodology for a new approach to imagery. It represents a synthesis of many concepts, ideas, and procedures developed within the traditional paradigms, while at the same time avoiding the apparent shortcomings of these other approaches. Fundamental issues neglected by the other systems concerning imagery's role in psychophysiology, psychosomatics, and psychopathology will then be addressed.

I. PARADIGM ONE: EXPERIMENTAL-COGNITIVE

1. Early Experimental Psychology: Structuralism

As an experimental science, psychology began in Germany with the researches of Fechner and Wundt and in England with Galton. The aims of experimental psychology were seen by these early investigators to be identical to those of the natural sciences: to make the phenomena of mental life the subject of investigation in the same way as those of the natural world are explored by the traditional sciences. Wundt described the operation thus:

"we must endeavor so to control our mental processes by means of objective stimulation of the external organs. . .that the disturbing influence which the condition of observation tends to exercise upon them is counteracted. This control is given by experiment. Not only does experiment, here as elsewhere, enable us to produce a phenomenon, and to regulate its conditions, at our pleasure: it possesses in psychology an especial importance, in that it alone renders self-observation possible during the course of a mental process" (Wundt, 1894; Ed. Robinson, 1977, p. 13).

Wundt's program was adopted by many distinguished students, none more enthusiastically than the Englishman Titchener, who established his laboratory at Cornell (Titchener, 1896). The school of "elementism" or "structuralism" evolved. Mental processes were to be analyzed into their basic elements and the projected goal was a kind of "mental chemistry" a la John Stuart Mill. As is well-known, the analysis of imagery and thinking by these early structuralists led to an unresolved controversy over "imageless-thought" (Humphrey, 1963; Woodworth, 1938, pp. 784-890). Although much of value was produced by the Wurzburgers and other introspectionists, personally I agree with Woodworth's (1938) conclusion that "the whole question may well be shelved as permanently debatable and insoluble" (p. 788). The debate over imageless-thought can be seen as part and parcel of a much larger issue: What limits are there in the applicability and validity of the methods of introspection? It is self-evident that introspective methods can only be validly applied to those mental processes which produce a conscious component or end result. Much of our mental activity is unconscious and therefore unavailable to introspection. Wundt was aware of this, and he was unimpressed by those who sought to discover the unconscious "determining tendencies" of imagery and thought using introspection. The early structuralists, however, were concerned more with the analysis of conscious experience by breaking it down to its basic components than they were with its synthesis and overall content.

2. Cognitive Psychology

The resurgence of research on imagery and cognition over the last quarter-century has followed many of the precepts of the early experimental psychology of Fechner and Wundt. The methodology - with one major exception - has also been similar, for example, the measurement of the time of the various components of mental tasks, the logging of psychophysical relationships for stimuli either present or absent and the establishment of the memory and problem-solving capabilities of the average normal adult human. This modern brand of experimental psychology, which we now all think of as "cognitive psychology," purports to provide a general analysis of how human beings "process information." An important tool, introspection, is hardly ever used by the modern-day experimentalist however, as it is generally regarded as unreliable.

An important critique of much of the best known laboratory research on imagery conducted within the cognitive paradigm has been presented in this volume by Yuille (see Chapter 8). Yuille's critique can be extended well beyond the imagery field: it casts a shadow over the whole cognitive enterprise. Many of the major findings on mental rotation (Shepard & Metzler, 1971), image scanning (Paivio, 1975), mental comparisons (Kosslyn, Ball, & Reiser, 1978), and verbal learning (Paivio, 1983), can be shown to be artefacts of the procedures used and much more ambiguous than the original investigators believed. Yuille's solution is for cognitive psychologists to leave the laboratory and study "real world" cognition, in situ.

Yuille's critique, but not his solution, is reminiscent of the early behaviorists' objections to the structuralists' laboratory method of introspection. Are we again to be led away from the laboratory experiment?

I doubt it. Is it the method that is wrong, or is it being badly used? The latter, I suspect. The method of introspection is one laboratory method, not mentioned by Yuille, which has already produced, and will continue to produce, data of considerable value to psychological science. However, it must be used carefully and appropriately, with all the controls and caution that can be brought to bear.

The major shortcomings of the experimental-cognition paradigm lie not in its methods but in its *theoretical objectives*. A method is never wrong, only wrongly used. Cognitive theory addresses the issue of information processing in the average normal adult human being. Influenced by computer science and technology, "cognitive science" hopes to model and understand human psychology from the perspective of information-processing machinery (computers). It ignores the double-requirement of psychology to provide an explanation for both behavior and consciousness. It ignores or "averages out" individual differences, a major interest of one of its founders, Galton. It fails to include either *motivation* or *affect* in its theorizing: there is no evidence that machines possess these characteristics. Rarely does it consider how the average normal adult human being arrived at that position or how "it" might develop some future change of status. Cognitive psychology has failed to produce a single theory of the function of imagery in human development. Its unrepentant analysis of the timing, storage, retrieval, and processing of all of the mental components of a small set of artificial laboratory tasks appears to be a symptom of a much more general and serious malaise: *each of the components of mental functioning, so carefully nursed in the cradle of a narrow laboratory experiment, normally operates within the subject's psychological system as a whole.* This is the most fundamental issue in the use of the laboratory experiment as a scientific tool within psychology. We do not learn much about the nature of an elephant by making a detailed examination of one of its toes! In Wundt's words: "Sense, feeling, idea, and will seemed to be related activities, and they appeared, further, to be bound together by the unity of self-consciousness" (Wundt, 1984; Ed. Robinson, 97, p. 8).

The integration of psychological events within the unity of consciousness is a mystery. It must always remain so, when seen through the myopic lens of a paradigm which aims to study "information processing" in a vacuum, in isolation from needs, affects, and human development.

II. PARADIGM TWO: PSYCHOANALYTIC-DISSOCIATIVE

The contribution of this approach to the study of imagery has been recently reviewed in this volume by Peter McKellar (see Chapter 2) and therefore I can afford to be brief. McKellar reminds us that "imagery" occurs in psychoanalytic writings under a large number of aliases, for example, in the process of repression, free association, dream recall, and transference. The autonomous nature of some imagery and the dissociative states which may occur during the more dramatic "eruptions into consciousness" have led many theorists to hypothesize dipsychic or polypsychic theories of personality (e.g., Janet, Freud, Jung, Prince, Murray). Methodology in this tradition has revolved around the problem of understanding the unconscious mechanisms presumed to underlie dreams, waking imagery, thinking and behavior. Methods have included free association, dream recall, projective tests, automatic writing, crystal gazing, and guided fantasy. The structure of the imagery is interpreted as an indicator of the subject's unconscious needs, conflicts, problems and concerns. Therapy aims to bring into consciousness the associative links between different areas of the subject's experience and behavior, enabling insights to be attained and conflicts to be resolved. McKellar suggests that "we view psychoanalysis as a technique for ridding the personality of some of the undesirable aspects of dissociation."

A major assumption of the psychoanalytic-dissociationist paradigm is that the contents of dreams and images are not as they appear, that symbolic interpretation is needed for proper analysis. Other approaches to imagery and dreams appear, to this author, to be more compatible with neuropsychological research and are to be preferred on the grounds of parsimony (see Chapter 9). Furthermore, there is no clear-cut evidence that dream content can be predicted from waking mental experience, or vice versa, and psychoanalytical methods appear to lack scientific validity at the present time.

III. PARADIGM THREE: BEHAVIORISM AND S-R PSYCHOLOGY

Of necessity, this section will be brief. The conception and birth of behaviorism is attributed to J.B. Watson's (1913) manifesto, *Psychology as the Behaviorist Views It*. Watson attempted to translate mentalist concepts into behavioral terms, so imagery and thinking became subvocal speech consisting of small movements of the larynx and tongue. Consciousness and the methods of observing it, introspection, were offically banished, although Jacobsen (1973) reports that his friend Lashley assured him that "when he and Watson would spend an evening together, working out the principles of behaviorism, much of the time would be devoted to introspection" (p. 14).

While the behaviorist theory that we think with our muscles may appear ridiculous (Humphrey, 1963, pp. 185-216), it at least draws our attention to some of the somatic components of conscious imagery, an aspect which Wundt had ignored. Imagery does have both its central and peripheral components, and behaviorist research has made a contribution to the measurement of the latter and, through learning theory, to the theoretical understanding of the former (see Chapters 3 and 4). It has alerted psychologists of all leanings and persuasions to the simple fact that most of what we do, think, feel and need is a function of our developmental history. However, behaviorism and S-R psychology's account of imagery must always remain incomplete because it ignores its most essential feature: consciousness.

Eysenck's (1983) recent statement on behaviorism, imagery and behavior therapy clarifies some of the issues involved from a neo-behaviorist viewpoint. Eysenck points out that according to Hull: "concepts are acceptable if they are empirically pinned down from both ends, i.e. if both the stimulus and response side can be adequately specified and quantified. In the case of imagery, the stimulus side can be quite rigorously defined in terms of instructions"(p.30).This position is identical to that of cognitive psychology (e.g., Paivio, 1972). Eysenck (1983) goes on to say: "When it comes to the response side, it has been equally possible to pin down with some accuracy the effects of imagery. Instructions to use sexual imagery, for instance, can be shown to be effective by making suitable measurements of the degree of erection obtained in the male, or comparable measures of vaginal temperature and lubricity in the female" (p. 30). Clearly, the behaviorist position has come a long way from Watson's (1913) manifesto, and the use of imagery in the many varied forms of behavior therapy from Wolpe's (1958) systematic desensitization to Cautela's (1977) covert conditioning indicates a convergence of the cognitive and behaviorist paradigms.

IV. PARADIGM FOUR: NEUROPSYCHOLOGY AND PSYCHOBIOLOGY

This paradigm includes neuropsychology, physiological psychology, psychophysiology, and parts of neurophysiology, neurology, neuroanatomy, psychopharmacology, and neurosurgery. It may seem doubtful that the term "paradigm" in its usual sense can actually be applied to such a diverse group of disciplines, although they all fall under the general umbrella of "brain research." The common precept is the conviction that the final explanation for experience and behavior lies within the nervous system. Chapter 9 provides a review of neuropsychological research on imagery and so again I can afford to be brief

Psychophysiological researches have used a variety of measures and indicators of imagery including the EEG (Barrett & Ehrlichman, 1982), eye movements (Marks, 1973a), scanpath conistency (Marks, 1983a), pupillary reactions (Simpson, Molloy, Hale, & Climan, 1968), heart rate changes (Lang, Kozak, Miller, Levin, & McLean, 1980), respiration changes (Golla, Hutton, & Walter, 1943), electromyography (Jacobsen, 1931), saliva flow (White, 1978), flash electroretinograms (Kunzendorf, Brigell, & Peachey, 1984) and many others. All of these studies provide strong support for the notion that imagery has a somatic component along with the quasi-sensory and meaning components.

However, like the behavioral and cognitive approaches, the psychophysiological paradigm has failed to show much interest in the structure of different kinds of images and to relate this to the dependent variables; and, not surprisingly, it has shied away from a close examination of the introspective data upon which the scientific study of imagery and experience actually depends. While this paradigm is making tremendous progress in the discovery of mind-brain and brain-behavior relationships, it must be remembered that psychology cannot be reduced to physiology. The nature of many psychological questions are of intrinsically different kind and do not require, nor would they necessarily be helped by, neurophysiological answers.

V. PARADIGM V: PIAGETIAN DEVELOPMENT PSYCHOLOGY

Apart from the research to be described in the next section, Piaget and Inhelder's (1971) *L'Image Mentale chez L'Enfant* provides the only thorough investigation in the history of psychology of *image structure*:

"The normal adult is able to imagine static objects (a hexagon, a table, etc.), movements (e.g. the swinging of a pendulum, the accelerated downward motion of a moving body on an inclined plane), and known transformations (e.g. the dividing of a square into two rectangles). He is able, too, to anticipate in images transformations which are new to him - to anticipate, for instance, that when a square sheet of paper is twice folded into two equal parts and the point of intersection of the folds cut off, a single hole will be seen, whereas if it is folded *three* times into *two, two* holes will be seen" (Piaget & Inhelder, 1966; Ed. 1971, p. 1)

Piaget and Inhelder, unlike the behaviorists and early structuralists, did not conceptualize the image as a reproduction of perception but as "an active and internalized imitation" (Piaget & Inhelder, 1971, p. 3). So-called "reproductive images" (a misnomer) could be immediate (I) or deferred (II), and they could be "static," "kinetic" or "transformational." So-called "anticipatory images" could be "kinetic" and "transformational" but not "static." Transformational images of both the reproductive and anticipatory types could be of two basic forms: images of the "products" of the transformation, or of the "modification" itself.

The theoretical aims of the Geneva study of imagery were: (1) to determine the stage of development for each kind of image; (2) to determine whether image development is autonomous from or dependent upon the evolution of intelligence as a whole; (3) to study the relationship between the sensory and motor aspects of images; (4) to investigate the status of images as to their symbolic or realistic quality; (5) to investigate the potential role of imagery as a system of signification in thinking; (6) to investigate the role of images in the "figurative" and "operative" aspects of cognition; and (7) to examine the role of images in spatial or geometrical intuition.

The methodology of the Piagetian school is to present to subjects of various ages a stimulus and to require a selection from a number of options, or a drawing or gesture representing the stimulus after some displacement, rotation or other transformation. The results for subjects of different ages enable conclusions to be drawn concerning the nature and function of the imagery available at different stages of development.

Piaget and Inhelder observed two "decisive moments" in the evolution of imagery: (I) imagery's initial appearance in static form only at the same time as the formation of the symbolic function (1 1/2 to 2 years); (II) the emergence or rise of anticipatory images, that is, of kinetic and transformation images, at the operational level (7 to 8 years). Images are based not upon perceptual origins, in Piaget and Inhelder's view, but upon internalized imitation of objects and actions. Furthermore, images are symbolic and provide a semiotic function which complements language. Piaget and Inhelder give two reasons for the latter conclusion. First, the sins of language are always social and too abstract for personal convenience. Hence, one invariably finds personal images to supplement even the most abstract and uniform terms such as those denoting the number series, these images having been studied since the days of Galton. Second, images are required to evoke and think about what has been perceived (objects, events, personal relations, etc.) "so that it can be drawn on as required for multiple purposes of adaptation" (Piaget & Inhelder, 1971, pp. 380-381).

The following passage provides a summary of Piaget & Inhelder's view of imagery and knowledge. It seems so pertinent to the paradigm to be described in the next section as to warrant quotation:

. . . knowing the object means acting upon it in order to transform it, and discovering its properties through its transformations. The aim is always to get at the object, but also on the exchange or interaction between subject and object resulting from the action and reaction of the two. As action takes effect two kinds of acquisition ensue. The first relates to the properties of the object, which have still to be interconnected, since they at first appear unorganized. The second concerns the co-ordinations of action itself, which require to be structured, for they are not performed.... It is impossible, therefore, to consider knowledge as a copy of the object, since it presupposes the intervention of a general co-ordination of action, whose stuctures are the sole means of interpreting the objective datum (Piaget & Inhelder, 1971, pp. 387-388).

VI. PARADIGM SIX: NEW STRUCTURALISM

1. Rationale, Precepts and Objectives

The perspective provided by the five paradigms described has given us an incomplete picture of imagery processes and their relationship to psychological functioning as a whole. A number of fundamental issues which seem to represent essential feature of image processes have not been properly addressed. These issues may be stated as follows: (1) What is the nature, structure, and function of imagery in the context of human development? (2) What is the relationship between imagery and other fundamental processes: learning, thinking, emotion, perception, and the senses? (3) What is the relationship between images and somatic processes occurring in healthy and unhealthy states of the organism? (4) What is the relationship between imagery structures and psychopathology? (5) What mechanisms underlie the effective application of imagery therapy for the control of mental and physical disorders including the so-called "psychosomatic" disorders?

In my view, we must construct a new paradigm for psychology and imagery theory. Instead of illuminating the essential issues and important questions, traditional paradigms seem to have obfuscated them. In adopting the methods of neurology, neurophysiology, physics, and computer science, psychology cannot realistically hope to advance its knowledge of its core area, the "psycho" or mind. Psychology should not be afraid to construct new methodologies and paradigms for advancing its aims and objectives. In its search for recognition and respectability among more established sciences, "behavior science" risks becoming a white-coated scientific pretension with the trappings of a science and its methodology but no real discoveries, advances in understanding, or laws comparable to those in other sciences, and preciously few reliable facts. A new paradigm is required which starts from the beginning, acknowledging what has come before, but which challenges the assumptions and conclusions previously reached.

One of the fundamental problems with psychology's approach to imagery has been its unwillingness to accept introspectively obtained data on the contents and structure of conscious experience. The earlier structuralism of Wundt, Titchener and "the Wurzburgers" failed, not because introspective methods are invalid, but as a result of their inappropriate use. The early structuralists, particularly those working in Wurzburg, tried to apply these methods to areas of mental functioning which are conducted unconsciously. There is a special category of mental activity which can be examined at an introspective level, since what is observed is always the product, not the processes themselves. Images are in this special region of introspectively-available data. But researchers following the behaviorist and cognitive paradigms have rejected or made hardly any use of this valuable data-source. Psychoanalytically-inclined investigators have analyzed such data, but the procedures for analysis are time-consuming and subjective and appear to have little predictive value. A new paradigm for the psychology of imagery would be wise not to repeat these errors.

The New Structural approach attempts to answer questions neglected by other paradigms by providing an account of the nature, structure and function of imagery in both "normal" and "abnormal" states of the organism. The paradigm adopts the philosophical stance of identity theory and assumes that images are neither purely mental nor purely physical, but that they are both. The structure of the image, including the experiential, somatic, and semantic features, contains the essential information on the individual's psychological development. The structural theory of image formation assumes Ahsen's (1977, 1984) Triple Code Model of images, which take the form of (i) a quasi-perceptual experience occurring in consciousness: image (I); (ii) a psychophysiological component: somatic response (S); and (iii) an interpretation, or meaning in the verbal-semantic system: meaning (M). The triple 'ISM" code is normally present for all imagery, all the time. Images are never free of associated emotions and thoughts and like movements of the body, they are, at the same moment, both a stimulus and a response (see Chapter 1).

The structure and content of images are determined by the individual's unique developmental history. Each image represents the "essence" of a set of past experiences as the imager constructs them. Every image is therefore a product of both memory and the imagination. Traditional classifications of imagery experience have attempted to pigeon-hole images into four categories: after-image, eidetic image, imagination, and memory (Richardson, 1969). This classification system is artificial and conceals as much as it reveals about the nature of imaging processes. How the "essence" is distilled from the rough and inconsistent malt of life's experiences is one of psychology's most neglected theoretical questions. One of the major goals of the structural theory of images is to provide this account of image formation. Stated more specifically, imagery theory must provide descriptions, predictions, and explanations of image structure and content in the context of discernible facts concerning the individual's developmental history and current state of consciousness.

2. Imagery, Conflict and Stress

An organism experiences conflict when competing or opposite emotions are set off at the same time or in quick succession. The somatic components of emotional reactions involve the two branches of the autonomic nervous system, while the central nervous system acts as an appraiser or executive monitor. Conflict can be said to be acutely stressful and repeated exposure to conflict situations is chronically stressful. All emotional events and situations, are recorded in the central nervous system as eidetic images. Eidetic images are universal and biologically ancient mechanisms for representing experience (see Chapter 9). Repetitions of a particular conflict scenario lead to the establishment of highly structured and inflexible images. When images are eidetically experienced, emotions similar to those which occurred during their establishment are also experienced. The repetition of conflict situations therefore has two consequences: (1) the development of chronic stress; (2) the development of rigidly structured images.

The most potent sources of emotionality in infancy are the baby's parents, initially the mother. When conflicts of any kind occur between the baby's and the mother's needs, the baby will be exposed to stress. As the child grows older and comes to acknowledge both parents, the mother and the father, there is the potential for further sources of conflict: between the parents, who may have conflicting needs at any one time, and between the child and each parent. These sources of conflict, if repeated conflicts occur over a number of occasions, will give rise to stress and inflexible images in the eidetic process. The various components of the conflict situation will be represented somatically by the images in similar form to the emotions originally experienced in the conflict situations. Psychosomatic symptoms are the final result of chronic levels of stress resulting from emotional conflict.

The mechanisms by which images are acquired follow the principles of learning, namely, the association of neural processes which are set off by each of the component stimuli (see Chapters 3, 4 and 9). The structure of the images therefore reflects the structure of the components of the situations associated with image formation. However, the images are not simply copies or *reproductions* of any of the situations which cause the images to be formed. This is a fundamental principle of image formation and is the source of much confusion in the literature (see Marks & McKellar, 1982, for a more complete discussion of this point, and also Ahsen, 1977, 1984, and Humphrey, 1963, Chapter 9).

Eidetic images are condensations or distillations of the essential features of the causative stimuli. In other words, they are constructions *not* reproductions. The principles of generalization from learning theory apply, and only the common features of all the situations which have been grouped together are retained in the eidetic. Hence, any particular eidetic is a construction built up from many situations and represents a summary of the essential features of all of the situations so grouped, not any one of them in particular. Images of important stimulus objects such as one's parents therefore convey a considerable amount of information: they represent the nervous system's highly edited summary of many of the most relevant experiences. Unusual images reflect unusual experiences. Stressful images reflect stressful experiences. The structure of the image tells us much of what we need to know about the development and psychological history of the imager.

3. Methodological Issues

This new paradigm is so named because it is based upon a structuralist conception of psychology with certain important modifications. Furthermore, it has taken over elements from all of the other paradigms, producing a synthesis which, although overlapping, is different from all in at least one major characteristic.

New structural methodology differs radically from that of other paradigms because it unashamedly sets out to obtain as complete a phenomenological report as possible of the subject's imagery which is produced following standard instructions. The method of introspection is used because it is the only method which provides the necessary information. The results obtained indicate that introspection provides a highly reliable and valid indicator of imagery quality, content, and structure. The results on imagery quality, for example, do not depend upon aretfacts such as demand characteristics and the procedural variables described by Yuille (Chapter 8) and other critics (Chapter 1) because the subject cannot possibly monitor or control the variables upon which his/her image quality can be shown to depend. These variables are all objectively measured and have values which are unknown to the individual subject and which are unalterable through personal volition.

Consider, for example, visual image vividness. A multitude of studies conducted in my laboratory and elsewhere testify to the validity of introspective reports of imagery vividness. I will cite only three. First is the study by Hodgson (1977) who obtained data on a group of 30 university students with respect to the following variables pertaining to their perception and recall of some pictures: total picture recall, maximum saccadic distance during picture perception, average fixation rate during perception, average fixation rate during imagery, average blink rate, and average image vividness prior to each recall. The multiple correlaton coefficient of image vividness with these five variables was $+0.80$, giving a coefficient of determination of 0.64. Hence, 64% of the variance in the subject's image vividness reports was explained on the basis of these five predictor variables. Are we to believe these subjects knew what the investigator was expecting and discovered the other subjects' scores (using some form of telepathy and precognition) and adjusted their behavior accordingly?

Molteno (1982) recorded the heart rates of students who visualized a number of pleasant and unpleasant scenes. Half the subjects were high scorers on the *Vividness of Visual Imagery Questionnaire* (VVIQ: Marks, 1973b) and half were low scorers. The heart rate change from a baseline (neutral imagery) condition is given in Table 1.

Table 1

Heart rate change for good and poor visualizers for four affect zones (beats/minute)

Visualizer Group	Highly Negative	Negative	Positive	Highly Positive
Good	4.0	2.5	2.9	4.4
Poor	0.9	1.2	1.5	2.7

Subjects reporting vivid imagery showed a larger heart-rate response than subjects reporting vague, ill-defined imagery. Again, are we to believe that these subjects willfully altered their heart rate to please the experimenter? On what information could such alteration be rationally based?

Isaac (1984) asked a group of internationally-ranked trampolists who represent their countries in this event to fill out the VVIQ and a similar questionnaire designed to assess movement imagery (VMIQ). The correlation between the VVIQ and their international ranking was 0.40 (p <.05) and between the VMIQ and ranking was 0.52 (p<.05). Are we to assume that these athletes adjusted the true values of their imagery vividness ratings using some telepathic knowledge of the values given by the other members of the sample? Or is there a simpler, more rational explanation, that vivid imagery could well be of benefit in the acquisition and competition performance of a complex motor skill? Many more experiments of this kind have been conducted, and all the results point in exactly the same direction: verbal reports of imagery provide highly reliable and valid predictors of mental functioning. More details of this research have been published elsewhere (Marks, 1972, 1977, 1983a, 1983b).

VII. IMAGERY AND PSYCHOPATHOLOGY

1. Overall Image Quality

The New Structural paradigm's primary goal is to provide a scientific account of the structure and content of human consciousness. In the field of imagery the New Structuralist aims to investigate as fully as possible the structure and content of images, more particularly, significant images. This analysis of imagery is then related to the subject's developmental history and projected future.

This section examines the relationship between imagery and the development of psychopathology. Do individual differences in imagery vividness and control have any correlation with psychopathology? Or, as the New Structuralist theory would predict, must we examine the structure of the images rather than their overall quality? These questions have prompted two recent studies in my laboratory.

The first study tested groups of university students who were either strong or weak visualizers on a number of measures of their personality (Power, 1984). The basic aim of this project was to determine whether students who displayed excellent eidetic ability showed more evidence of prepsychotic tendencies or other personality differences than a group of weakly eidetic imagers.

The subjects were selected using a composite score based upon two eidetic imagery tests: the Open Circle Test and the Eidetic Picture Test (Marks & McKellar, 1982). Both tests have a maximum score of 8. Mean scores and standard deviations for the two groups of subjects on the two tests of eidetic ability are given in Table 2.

The tests included a compilation of three scales designed by Chapman, Chapman, and Miller (1982): (1) The Physical Anhedonic Scale, consists of 61 items assessing a lowered ability to experience

Table 2
Mean scores of strong and weak eidetikers on the Open Circle and Eidetic Picture Tests

Test	Weak Eidetikers		Strong Eidetikers	
	Means	S.D.	Means	S.D.
Eidetic Picture	1.05	0.99	7.85	0.37
Open Circle	0.45	0.76	6.95	0.69

pleasure, for example, "On hearing a good story, I have wanted to sing along with it." (2) The Perceptual Aberration Scale, consisting of 35 items assessing perceptual aberrations including distortion of the body image, for example, "I have sometimes had the feeling that one of my arms or legs is disconnected from the rest of my body." (3) The Non-Conformity Scale, consisting of 51 items assessing impulsivity and lack of regard for ethical and social moves, for example, "I find it difficult to remain composed when I get into an argument."

In addition to the Chapmans' tests, Power administered an auditory detection task modeled closely on the measure employed by Mintz and Alpert (1972). Each subject was played a series of 24 recorded phrases along ith varying levels of background noise. The subject was required to repeat exactly what (s)he heard and also give a confidence rating along a 6-point scale. For each item, an accuracy score was calculated using the formula:

$$\text{Accuracy} = \frac{\text{Number of syllables correct} \times 100}{\text{Number of syllables in sentence}}$$

Pearson's product-moment correlation was calculated between the 24 accuracy and confidence scores for each individual subject. This correlation provides an index of the subject's overall tendency to experience auditory hallucinations. A high correlation indicates that the subject adjusts his or her confidence ratings according to the veridicality of the report. A low or negative correlation indicates that the subject tends to report inaccurate phrases with overly high confidence.

A further personality measure used by Power was the Embedded Figures Test (Witkin, Ottman, Raskin, & Karp, 1971). The subject is required to locate a previously seen simple figure within a more complex geometrical design, which provides a measure of the subject's ability to differentiate figure from ground. The Embedded Figures Test purports to assess boundary confusion and the level of field dependence or independence.

The results obtained by Power are given in Table 3. Not a single statistically discernible difference occurred between the two groups of subjects. There was no evidence of any personality differences indicative of prepsychotic tendencies between strongly and weakly eidetic subjects.

Table 3
Mean scores of weak and strong eidetikers on the personality tests

Test	Weak Eidetikers		Strong Eidetikers	
	Means	*S.D.*	*Means*	*S.D.*
Embedded Figures	41.7	29.4	37.3	20.9
Auditory Detection	(r =	0.70)[1]	(r =	0.35)[1]
Physical Anhedonia	8.80	5.9	12.5	19.7
Perceptual Aberration	8.65	9.3	11.0	9.0
Non-Conformity	16.40	9.2	15.6	8.1

[1] See text for explanation of this measure

We can conclude from this research that eidetic imagery appears to represent a normal ability of healthy individuals (Ahsen, 1977; Marks & McKellar, 1982). Subjects with vivid eidetic images are no more likely to display prepsychotic features or other personality disturbances than subjects whose eidetic imagery is characteristically less vivid. This finding supports the observation of Wilson and Barber (1983) who found that highly fantasy-prone personalities, characterized by strong eidetic imagery and high hypnotizability, are not more likely to develop neurotic or psychotic disorders. We are led toward the prediction that it is *image structure* which will show a strong relationship to psychopathology rather than gross features such as overall vividness or eidetic quality.

2. Structural Features of Imagery

The second study conducted by Molteno (1984) tested a group of university students on the Eidetic Parents Test (Ahsen, 1972). The aim of this project was to check certain assumptions in the structural theory of image formation concerning the relationship between image contents and mental health. Significant events in an individual's development are registered as an eidetic image with all three of the components (I, S, M). Eidetic images may become fragmented and some of the components may become dissociated over the passage of time. Images of the parents and of any conflict or disturbances between the parents or between the child and the parents are held to be especially important in uncovering the basis for psychological and psychosomatic problems. The Eidetic Parents Test (EPT; Ahsen, 1972, 1977) was specifically designed to investigate these relationships, and aims to provide the subject with therapeutic insight into the sources of conflict and disorder. The full version of the EPT contains 30 items. In this study a modified, shortened version containing only scenes was used. A brief, highly edited summary of the eight scenes follows. A complete account of this study will be published elsewhere.

(1) Picture your parents in the house where you lived most of the time with them, the house that gives you a feeling of being a home. Notice where your parents are and what they are doing and how you feel when you see them.

Q. Where are they?

Q. What are they doing?

Q. How do you feel when you see the images?

(2). Picture your parents standing directly in front of you.

Q. Who is on your left and who is on your right?

Q. Can you reverse the figures?

Q. Do you experience any difficulty when you try to do so?

Q. Do the images spontaneously return to their original position or
remain in the new position?

(3) Item 2 repeated.

(4) Item 2 again repeated.

Q. Who is on your left and who is on your right?

Q. Who is nearer and who is further away from you?

(5) Item 2 again repeated. Then: Now bring your mother's image closer to you. Notice whether you are able to do it easily or with difficulty. Notice what you feel as you bring her closer to you and how your father feels as you bring her closer.

(6) Item 2 again repeated. Then the subject is asked to bring the father closer and notice the resulting feelings in self and the mother.

(7) See the father close and bring the mother closer to the father. Notice the accompanying feelings.

(8) See the mother close and bring the father closer to the mother. Notice the accompanying feelings.

Forty-five subjects (21 male, 24 female) participated in the individual administration of the above eight items while recordings of their heart rate were obtained. The Delusions-Symtoms-State Inventory (DSSI) was administered to all subjects. The DSSI is self-report checklist covering a wide range of symptoms and is designed to aid psychiatric diagnosis. The 84 true-false items are divided into 12 syndromes which form four hierarchical classes of illness: Neurotic States, Neurotic Symptoms, Integrated Psychosis, and Disintegrated Psychosis. As score of 4 or more on any set of items is considered significant.

The study confirmed a number of predictions of the structural theory of imagery. Specifically, approximately two-thirds of the subjects (64%) imaged the father on the left in response to Item 2, replicating the findings of Ahsen (1977). A significantly greater proportion of those "seeing" their father on the right had significant symptoms (50%) compared to those "seeing" their father on the left (24% only). Also the individuals in the former group reported more severe and wide-ranging symptoms. Fourteen of 15 students with significant symptoms "saw" the father or mother nearer (rather than seeing them as equidistant) while only 20 out of 30 with no symptoms "saw" the parents apart. An interaction also occurred between the father's position and relative parental distance: *..those who 'saw' their father on the right had much higher psycho-pathology scores than those seeing their father on the left but only when one or the other parent was 'seen' as closer"* (Molteno, 1984, p. 55). These results are shown in Table 4.

Table 4
Average scores on the Delusion-Symptoms-States Inventory as a function of father position and relative parental distance

Father Position	Relative Parental Distance					
	Father Nearer		Equidistant		Mother Nearer	
	Means	S.D.	Means	S.D.	Means	S.D.
Right	15.5 n = 6	12.7	1.3 n = 3	0.6	16.7 n = 7	13.3
Left	5.4 n = 7	4.7	4.4 n = 8	3.8	4.2 n = 14	3.9

The heart rate data were charted as changes from baseline recorded in the relaxation periods which occurred between consecutive parental images. A significant interaction occurred between the father's position and relative parental distance. Compared to subjects with their father on the left side, subjects who "saw" their father on the right showed a lesser increase in heart rate if the parents were equidistant, a greater increase if the father was nearer, and an approximately equal increase if the mother was nearer. The somatic component of parental images is affected by the same interaction of structural factors which was associated with differences in psychopathology. We predict that investigations of psycho-somatic disorders will also reveal singificant relationships between image structure and symptomatology.

Molteno's study, like Power's, found no significant differences in psychopathology scores between subject groups differing in overall imagery ability. Image content, not overall vividness or clarity, is correlated with the expression of symptoms.

The Molteno study of image structure is an important one as it confirms, under laboratory controlled conditions, that image content provides a highly useful indicator of the subject's degree of psychopathology. Further research along these lines is clearly warranted. We appear to be exploring an exciting and uncharted frontier of human consciousness. There is little doubt that there is much yet to be discovered about image structure and its transformations.

The New Structuralist paradigm is young and still developing. As it has borrowed its conceptual and methodological tools from elsewhere, it cannot claim in any sense (Kuhnian, or otherwise) to be revolutionary. However, it provides psychology with an argument to take a fresh look at a source of data which was long ago rejected for an insufficient cause. Consciousness and its many transformations are too important for psychology to ignore.

REFERENCES

Ahsen, A. (1965). *Eidetic psychotherapy: A short introduction.* Lahore: Nai Matbooat.

Ahsen, A. (1972). *Eidetic Parents Test and analysis.* New York: Brandon House.

Ahsen, A. (1977). Eidetics: An Overview. *Journal of Mental Imagery, 1,* 5-38.

Ahsen, A. (1982). Imagery in perceputal learning and clinical application. *Journal of Mental Imagery, 6,* 157-186.

Ahsen, A. (1984). ISM: The triple code model for imagery and psychophysiology. *Journal of Mental Imagery, 8,* 1-41.

Ahsen, A. (1985). Image psychology and the empirical method. *Journal of Mental Imagery, 9,* 1-40.

Barrett, J., & Ehrlichman, H. (1982). Bilateral hemispheric alpha activity during visual imagery. *Neuropsychologia, 20,* 703-7-8.

Boring, E.G. (1957). *A history of experimental psychology.* New York: Appleton-Century-Crofts.

Bugelski, B.R. (1982). Learning and imagery. *Journal of Mental Imagery, 6,* 1-92.

Chapman, L.J., Chapman, J.P., & Miller, A.N. (1982). Reliabilities and Intercorrelations of Eight Measures of Proneness to Psychosis. *Journal of Consulting and Clinical Psychology, 50* (2), 187-195.

Eysenck, H.J. (1983). Behaviorism, imagery, and behavior therapy. *International Imagery Bulletin, 1,* 30-31.

Golla, F.L., Hutton, E.L., & Walter, W.G. (1943). The objective study of mental imagery. *Journal of Science, 89,* 216-223.

Hall, H. (1984). Imagery and Cancer. In A.A. Sheikh (Ed.), *Imagination and healing.* Farmingdale, NY: Baywood Publishing Co.

Hodgson, I. (1977). Visual imagery and eye movements. Honours dissertation, University of Otago, Dunedin, New Zealand.

Humphrey, G. (1963). *Thinking.* New York: John Wiley.

Isaac, A. (1984). *Kinesthesis, imagery and mental practice in complex skill acquisition.* M.Ph.Ed. dissertation, University of Otago, Dunedin, New Zealand.

Jacobsen, E. (1931). Variation of specific muscles contracting during imagination. *American Journal of Physiology, 96,* 115-121.

Jacobsen, E. (1973). Electrophysiology of mental activities and introduction to the psychological process of thinking. In F.J. McGuigan & R.A. Schoonover (Eds.). *The Psychophysiology of thinking.* New York: Academic Press.

Kosslyn, S.M., Ball, T.M., & Reiser, B.J. (1978). Visual images preserve metric information: Evidence from studies of image scanning. *Journal of Experimental Psychology: Human Perception and Peformance, 4,* 47-60.

Kunzendorf, R.G., Brigell, M., & Peachey, N.S. (1984). *Effects on visual imaging on flash electroretinograms.* Unpublished manuscript. University of Lowell, Massachusetts.

Lang, P.J., Kozak, M.J., Miller, G.A., Levin, D.N., & McLean, A., Jr. (1980). Emotional imagery: Conceptual structure and pattern of somato-visceral response. *Psychophysiology, 17,* 179-192.

Marks, D.F. (1972). Individual differences in the vividness of visual imagery and their effect on function. In P.W. Sheehan (Ed.), *The function and nature of imagery.* New York: Academic Press.

Marks, D.F. (1973a). Visual imagery differences and eye movements in the recall of pictures. *Perception and Psychophysics, 14,* 407-412.

Marks, D.F. (1973b). Visual imagery differences and the recall of pictures. *British Journal of Psychology, 64,* 17-24.

Marks, D.F. (1977). Imagery and consciousness: A theoretical review from an individual differences perspective. *Journal of Mental Imagery, 1,* 275-290.

Marks, D.F. (1983a). Mental imagery and consciousness: A theoretical review. In A.A. Sheikh (Ed.), *Imagery: Current theory, research and application.* New York: Wiley.

Marks, D.F. (1983b). In defense of imagery questionnaires. *Scandinavian Journal of Psychology, 24,* 243-246.

Marks, D.F., & McKellar, P. (1982). The nature and function of eidetic imagery. *Journal of Mental Imagery, 6,* 1-124.

Mintz, S., & Alpert, M. (1972). Imagery vividness, reality testing, and schizophrenic hallucinations. *Journal of Abnormal Psychology, 79,* 310-316.

Molteno, T.E.S. (1982). *Imagery: Heart rate responses to pleasant and unpleasant scenes.* Postgraduate Diploma of Science dissertation, University of Otago, Dunedin, New Zealand.

Molteno, T.E.S. (1984). *Imagery in the Eidetic Parents Test.* Master of Science dissertation, University of Otago, Dunedin, New Zealand.

Paivio, A. (1972). A Theoretical analysis of the role of imagery in learning and memory. In P.W. Sheehan (Ed.), *The function and nature of imagery.* New York: Academic Press.

Paivio, A. (1975). Perceptual comparisons through the mind's eye. *Memory and Cognition, 3,* 635-647.

Paivio, A. (1983). The empirical case for dual coding. In J.C. Yuille (Ed.), *Imagery, memory and cognition.* Hillsdale, NJ: Lawrence Erlbaum Associates.

Piaget, J., & Inhelder, B. (1971). *Mental imagery in the child.* London: Routledge & Kegan Paul.

Power, A. (1984). An investigation into personality correlates of eidetic ability. Master's thesis, University of Otago, Dunedin, New Zealand.

Richardson, A. (1969). *Mental imagery.* London: Routledge & Kegan Paul.

Robinson, D.N. (Ed.). (1977). *Lectures on human and animal psychology. Wilhelm Wundt (1894).* Washington, D.C.: University Publications of America.

Sheikh, A.A., & Jordan, C.S. (1981). Eidetic Psychotherapy. In R.J. Corsini (Ed.), *Handbook of innovative psychotherapies.* New York: Wiley.

Shepard, R.N., & Metzler, J. (1971). Mental rotation of three-dimensional objects. *Science, 171,* 701-703.

Simpson, H.M., Molloy, F.M., Hale, S.D., & Climan, M.H. (1968). Latency and magnitude of the pupillary response during an imagery task. *Psychonomic Science, 13,* 293-294.

Titchener, E.B. (1896). *A textbook of psychology.* New York: Macmillan.

Watson, J.B. (1913). Psychology as the behaviorist views it. *Psychological Review, 20,* 158-177.

White, K.D. (1978). Salivation: The significance of imagery in its voluntary control. *Psychophysiology, 15,* 196-203.

Wilson, S.C., & Barber, T.X. (1978). The Creative Imagination Scale as a measure of hypnotic responsiveness: Applications to experimental and clinical hypnosis. *Journal of Clinical Hypnosis, 20,* 235-249.

Wilson, S.C., & Barber, T.X. (1983). The fantasy-prone personality. In A.A. Sheikh (Ed.), *Imagery: Current theory, research, and applications.* New York: Wiley.

Witkin, H.A., Ottman, P.K., Raskin, E., & Karp, S.A. (1971). *A manual for the Embedded Figures Tests.* CA: Consulting Psychologists Press, Inc.

Wolpe, J. (1958). *Psychotherapy by reciprocal inhibition.* Stanford: Stanford University Press.

Woodworth, R.S. (1938). *Experimental psychology.* New York: Henry Holt.

INDEX